Pre Big Bang – Big Bang – After Big Bang

Flavien Phanzu Mwaka

Pre Big Bang – Big Bang – After Big Bang

The Verses of the Scholars – Volume III

Digital ISBN : 979-10-262-7701-9

Paper EAN : 9791026277026

Librinova

Email : contact@librinova.com
Internet : www.librinova.com

(Revised synthesis of Volume I and Volume II)

DEDICATION

This book is dedicated to all those who have supported me during the most difficult moment of my life including my aunt KIADI Pauline and her wife Dr. NZUZI TSHAKALA, my aunt KIADI PELAGIE and her husband Pastor PHAMBU BABAKA, my uncle JOSE NDUBUDI and his wife Julie, my cousin JP KIADI and his wife, my niece Emilie Ntoto, my friend Patrick Nsangu and his wife Maguy, to Anasthasie MBOSHA KIPOTE and Eugenie VELA. I do not forget also my assistants Grace KANZA TSHIEFU and Ruffine KADIAY LUKUSA.

May he find through this dedication the expression of my love, my gratitude and my thanks through all they have done for me.

Epigraphy

" In fact with the Big-bang model, we have, perhaps reinvented Genesis in scientific terms. We would have translated the Bible into the language in form of today's physics".

Aurélien Barrau : Big-bang and beyond", ed. Dunod, Paris 2019, P.56

Preface

Good research works with undeniable religious, scientific and cultural depth. The field covered is tough and defensible in the light of reliable material evidence.

This text should be published as a trial balloon for readers unfamiliar with cosmological material combined with biblical evidence.

I would like the publication to be partial (3 or 4 volumes) in successive editions and not all at once because of the rigidity of the addressed material.

Once again thank you and good luck!

Dean of the Faculty of Arts National Pedagogical University (UPN)/RDC.

Foreword

In the series of books "THE VERSETS OF SCHOLARS", we highlight the scientific equivalences of the names that were cited in literary form in the Bible.

This book is a follow-up to Volumes I and II of the book series "The Verses of the Scholars". Everything that is said here is like a symbiosis of these first two volumes. Why have we made such a revision? Because these first two volumes do not visibly reveal my greatest discoveries. When I was writing the press releases that did not bear any fruit, I thought, why not make visible what everyone is looking for, i.e. the contents of the sphere from Pre-Big-bang to big-bang?

These two contents are already developed in volume I, except that its title, which is very logical, namely : "Cosmology according to biblical and ancient models", suggests that we are dealing with a religious book with thoughts based on faith. Thus, the results of my research contained in Volume I may take some time before they are discovered. So to avoid such a situation, I felt it necessary to highlight what the scientific world really expects and to make it the title of this book.

Scientific literature is like mathematics. If you try to change sentences or words to avoid being accused of plagiarism, you risk tarnishing certain theories or the true course of certain phenomena. Should we refrain from using the word fusion because another author has used it? Not at all. Thus, to avoid the pitfalls of giving two similar explanations for the same phenomenon, my readers of

Volumes I and II will find that I have included certain paragraphs as such in this book. However, researchers know that with each new edition, there is always an improvement, either through the addition of a new idea or the improvement of a paragraph. This is what we have done in this volume.

In Volume I, which is entitled "Cosmology according to Biblical and Ancient Models", we have shown that all the great cosmological theories and the great sets of the Universe, which were the subject of discoveries in the 20th century, were already revealed and known in ancient times, especially the Pre-Big Bang, the Big Bang, the black hole, dark material and dark energy.

Volume I constitutes the basis of the Universal Cosmological Model, as we first explained all the biblical symbolisms and the points of the Main Divine linked to the Creation of the Universe. This took us through the first five chapters, which are likely to discourage readers who want to get to the point, before tackling the pre-Big-bang to the living in a detailed manner from chapter VI onwards.

In Volume II which is entitled "The Field of the Universal Cosmological Model", we have tackled several subjects on cosmology, presenting our point of view. We also highlighted the place of the Universal Cosmological Model in relation to other scientific models.

In Volume IV ; "From Noah to Barack Obama", in addition to explaining the scientific prowess achieved by certain prophets in the Old Testament, we have highlighted the signs that show that the time of the establishment of the Kingdom of Heaven on Earth was already fulfilled, as indicated in the Mayan calendar, and that humanity must expect new floods of purification.

Mt 24:38 *For in those days before the flood they were eating and drinking, marrying and giving in marriage, until the day*

Noah boarded the ark. 39 They didn't know until the flood came and swept them all away. This is the way the coming of the Son of Man will be.

Without going back to the previous volumes, where we explained all the symbolisms and parables that obscured the story of biblical creation, here we tackle the multiple questions that people ask themselves from the Pre-Big-bang to the Big-bang. Being the first to talk about them with such conviction, I felt it was necessary to further develop these concepts which have just come to light at the beginning of the 21st century with the advent of the Universal Cosmological Model. Namely, the 20th century was characterized by the enrichment of the Big- bang theory known as the Standard Model.

In his time, the Belgian scientist, Father Georges Lemaitre, one of the founding fathers of the Big-bang theory, said : "*A truly complete cosmogony should explain atoms as well as suns*". With the Universal Cosmological Model, we can say that this prediction has just been fulfilled.

In several of my assertions, I often mention the expression "according to scientists" instead of quoting the source directly. This simply means that it is an assertion that is already approved by science and taught in astronomy.

My great contribution to the Universal Cosmological Model is to have clearly explained the words, names and phrases that made the biblical scheme of creation look like a magician's game with phrases like "God said, God made, God created".

This scientific restitution of the biblical creation pattern in the Universal Cosmological Model transfers the biblical creation from cosmogony to cosmology. And this constitutes a new revolution in the development of scientific thought.

The Universal Cosmological Model is not a question of belief, God or religion. It is the scientific restitution of the first Cosmological Model discovered in the papyrus and whose author attributed all the effects to divine intervention, that is, to God's will. This primitive model, called the Scheme of Creation, is contained in the Bible, whether a scientist is a believer or not, his development has nothing to do with faith. This is what we have achieved in the Universal Cosmological Model.

The Author

Introduction

In this volume we will focus on the Pre-Big-bang to the Big-bang. The after Big-bang is already the subject of current science and is developed in the standard model. However, in Volume I, we have reconstructed the entire biblical scheme of creation, i.e. from the Pre-Big-bang to the living beings.

For a better understanding of the themes dealt with in this book, an explanation of the concept of biblical time is necessary in order to prepare readers who have not yet read Volume I.

It is in the Bible and not in science that I made the discovery of the pre-Big- bang and the phenomena that took place. However, Moses wrote it for the people of his time at 1500 BC. At that time, chemistry and physics were not as developed as they are today, nor were scientific vocabularies. Moses could only use terms that these people could understand. If we too understand him as such, it is because we reduce our intelligence to their level.

It is this update of Moses' language in our time and the restitution of his thought that is called the Universal Cosmological Model. Thus, it is in the chapter dealing with the Universal Cosmological Model, that we will explain in a succinct manner the symbolisms and parables that have obscured the history of biblical creation.

Thus, in the Universal Cosmological Model, which is the restitution of the biblical scheme of creation, we have detected three stages : an empty universe with total darkness (Gen 1 :1), then came the manifestation of God's power over water (Gen 1 :2), known under the scientific name ; the action of energies on original material. Through the First Universal Force, (super gravitation according to science), this power of God will lead to the third stage, namely, extreme warming generating rays. This is what is described biblically by the phrase : "God said, let there be light, and there was light" (Gen 1 :3).

This is how the whole dark period from the Pre-Big Bang until the appearance of the first light, according to the Bible, was described as the first day (Gen 1 :2-4). The Big-bang itself occupied the second period, known biblically as the 'separation of the waters from the waters' (Gen 1 :6-8) and the period of star creation or the post-Big-bang period occupied the third and fourth biblical days (Gen 1 :9 and 14).

These are the great discoveries that no scholar has ever known before, and which we will dwell on in the following pages.

Chapter I :
The Universal Cosmological Model

Section 1 : Presentation

1.1. Definition of the Universal Cosmological Model

The Universal Cosmological Model is the translation into scientific language of the scheme of creation, as written in the first and second chapters of Genesis. This scientific translation will henceforth be the foundation of all Biblical Models.

The Universal Cosmological Model, we have said, is the starting point from which everyone can research and draw comments and conclusions. That is to say that nobody, not even I, has a monopoly on the analysis of the Biblical Model. The latter is the set of all the Models that support the biblical pattern on the creation of the Universe. What I have developed and designed is called "The Universal Cosmological Model". Everyone can be inspired and enriched by this Model.

1.2 Contents of the Universal Cosmological Model

The Universal Cosmological Model sheds light on the history of the evolution of the Universe according to the biblical scheme of creation. I say evolution, not creation, because what comes beyond the Uncertainty Principle (energy-particle relationship) is difficult to demonstrate; that is, the transition from the spiritual (as explained in the Principle of Creation, a branch of the book of the Divine Principle by Rev. Dr. Sun Myung Moon) to energy in a scientific manner.

The originality of the Universal Cosmological Model compared to other scientific models in cosmology lies in the fact that it speaks from about the Pre- Big-bang, through the Big-bang to the appearance of man on earth in a harmonious and coherent way. Since the other Scientific Models only deal with one part of the cosmology, we can say that they are all embedded in the Universal Cosmological Model. "Embedded" means that it captures them and puts each one in its place to allow scientists to present us with a coherent scheme and the theory of Everything.

The Universal Cosmological Model is the scientific restitution of the biblical creation scheme. Its scope covers and studies events from the moment the word Universe was first uttered in the Bible through its definition of a "formless and empty earth" (Ge1:2), i.e. something that still had no material existence, otherwise it would never be formless and empty, until the creation of man. This study is limited to the level of origins and also attempts to unravel the mystery of the "why".

1.3. The branches of the Universal Cosmological Model

Building a cosmological model and having astrophysicists give you credit is a certain specialization. The specialization is not mining, but Moses, a Jewish scholar in ancient times. So the Universal Cosmological Model is part of the Jewish wisdom of ancient times. Perhaps I am only a self-taught exegete. But how, before me, has no one managed to reconstruct the biblical creation story in a scientific manner? The first reason is that I was the first to ask myself why the biblical creation story was thrown away by scientists.

In order to answer this question myself, in addition to my passion for science, I also had to have a body of knowledge that no one had before the books in the series

"The Verses of the Scholars" came out. This knowledge acquired in a personal (self-taught) way that people can join me through :

– The assiduous reading of the Bible

– The reading of other so-called sacred books

– Reading esoteric books

– The assiduous reading of the Divine Principle by Rev. Dr. Sun Myung Moon

– The participation in several educational seminars of the Unification Church of Rev. Dr. Sun Myung Moon, which became the Holy Family of the Heavenly Parent

– The assiduous reading of articles dealing with cosmological phenomena in scientific sites such as Futura-Sciences, Trust My Science, etc...

– The taking of notes by following scientific documentaries as a student following a course.

– The training on Astronomy Master at the Observatory of the Baronnies Provençales, in France, which makes me an amateur astronomer.

It is all this knowledge that has made me the first person in the Universe to be able to reconstruct the story of biblical creation. A story written 3. 500 years ago in papyrus found in the Dead Sea and transmitted as such in the Bible. This makes me the first researcher who discovered the horizon before the Big-bang and its contents up to the Big-bang.

I had said in Volume I that my restitution leading to the discovery of the Big-bang in the Bible surprised God himself, so much so that he came to give me the injunction to popularize it and to tell scholars that they should all be inspired by the biblical scheme of creation in their research. That is all God gave me as an injunction in a 2012 dream-vision, and let me make further discoveries myself to make sure that the biblical creation story is really true.

So said, the three main branches for me to be able to set

the Universal Cosmological Model are the knowledge of the main lines of cosmology, the biblical knowledge and the Principles of Creation of Rev. Dr. Sun Myung Moon. These three branches together form what I call astrotheology and must now be included in all faculties of theology, physics and philosophy.

Look, when we say, God created the Universe, we are in religion. But when we ask ourselves the question : "With what did God create the Universe ? We enter directly into cosmology and philosophy. Those who want to remain in religion say "everything was created by the Word" (Jn 1 :1-3). But Moses, who wrote the biblical scheme of creation, was the first to enter cosmology, saying 'man was formed from the dust of the earth' (Gen 2 :7) to answer the question : 'with what did God create man ?

(Working session at the Observatory of Baronnies Provençales -OBP- in France/September 2019).

Flavien PHANZU MWAKA in front of the dome of the Observatory of the Baronnies Provençales (OBP) in France. (September 2019)

Practical session inside the dome of the Observatory of the Baronnies Provençales (OBP) in France. (September 2019)

Some interns in practice inside the dome of the Observatory of the Baronnies Provençales (OBP) in France. (September 2019)

Moreover, the whole story of Genesis chapter 2 is based on cosmology. So, I no longer see why believers and scientists will still have contempt for reading Volume I of my book which is entitled "Cosmology according to biblical and ancient models". Scientists mistrust it because of the word "biblical" and believers mistrust it because of the word "cosmology".

Thus, in the first five chapters of Volume I, we wanted to explain to scientists the languages used in the Bible to which they were not accustomed. In the same way, we have also taken the opportunity to bring the ordinary man to understand scientific language and to lead believers to believe in facts instead of being confined to the chronology of creation according to the Bible with a 6,000- year-old Universe. As a result, the main subject of Volume I began in Chapter VI.

This way of proceeding, which we felt was necessary, earned us a lot of criticism and some even went so far as to call us fanatics. For us, the goal is achieved because all the astrophysicists who have read me have bought Bibles to check whether I was not myself inventing the verses, and believers are now beginning to take an interest in cosmology, knowing that they now have a say.

Just as everything that astrophysicists observe is part of the effects, which are, moreover, the object of their science, they must include in their discipline, as complementary or personal knowledge, the cause. And the cause is indeed the teaching of the Principle of Creation by Rev. Dr. Sun Myung Moon, as the philosophical basis of their teaching or knowledge. It is only through this teaching that they will be able to solve many enigmas. This concern responds to the words of Jesus who said to his disciples :

"I speak to you of earthly things and you do not believe, how shall you believe when I speak to you of heavenly things ?" (Jn 3 :12).

This promise of Jesus is fulfilled to this day by the Moon couple in their book "The Divine Principle", which contains the Principles of Creation.

Section 2 : The Basic Statement of the Universal Cosmological Model

2.1. The biblical scheme of creation

Indeed, no theologian up to that point has had the idea of thoughts beyond the biblical writings. Thus, the verses below remind us of the different phases of the history of the Universe according to the Bible, also confirmed by science.

1) Ge 1 : 1-2 "In the beginning God created the heavens and the earth (the cosmos), *and the earth* (the universe) *was formless and empty".*

Ge 1:2 *"There was darkness on the surface of the deep".* (First deformation of the fabric of space-time in the axis of the universe).

Gen 1:2 *"The Spirit of God moved over the waters".* (Gravitation on the original matter before Big-bang in the abyss).

Ge 1:3 "*God said, 'Let there be light and there was light'*; (consequence of the fusion on the original matter or of the Spirit of God moving over the waters).

Gen 1:4 *"God separated the light from the darkness"*; (scientific riddle to be solved).

Gen 1:5 "*And there was evening and there was morning, the first day*" (end of the before Big-bang).

2) Ge 1:7 "the separation of waters from waters" or Big-bang, according to science

(1) and the formation of the expanse (expansion of the universe).

Gen 1:8 "And there was evening and there was morning, the second day"; (the zero moment of our Universe).

3) Gen 1:9 "*Out of the waters below* (1) *the dry* (2) (or earth) *appeared*"; (beginning of the period after Big-bang).

Gen 1:13 *And there was evening and there was morning, the third day.*

4) Gen 1:14 *"Out of the waters above* (1) *there were lights* (2)".

Gen 1:19 *And there was evening and there was morning, the fourth day.*

5) Gen 1:20 *"And God said, Let the waters bring forth abundantly living creatures, and let birds fly over the earth toward the expanse of heaven.*

Gen 1:23 *And there was evening and there was morning, the fifth day.*

6) Gen 1:24*" God said, "Let the earth bring forth live animals after their kind,...*

Gen 1:27 *"God created man in his own image, in the image of God he created him; male and female he created them".*

Gen 1:31 *"And God saw all that he had made, and behold, it was very good. And there was evening and there was morning, and the sixth day.*

2.2. Literal restitution of the biblical scheme of creation

Thus, the Universal Cosmological Model, on which will be based all the analyses that will follow on the part of the different scholars and philosophers, will have as its basis the statement below.

The Universe was formless and empty. Given, on the one hand, an energy defined as darkness by the Bible in the abyss, and on the other hand, the spirit of God or a wind from God on the original material called "waters", there was the production of a radiation defined as light. From the beginning of the Universe in nothingness until the first light or radiation was produced, this is the period that the Bible calls 'day' (first day), a period that lasted for a time that could not be quantified until then.

The Bible further specified that this new energy or radiation had an effect on darkness. It came to counteract it, as well as the invisible material that was already at work in the early universe. This is expressed in the sentence "God separated the light from the darkness". And to show that these two energies would henceforth accompany the physical destiny of the Universe, God called darkness "night" and light or new radiation "day", for lack of the appropriate scientific words at that time.

The original material or "waters" according to the biblical term, continuing its warming cycle after the production of light (light be and light was), had a burst, an expansion or an expulsion into the abyss and that is what science calls "Big-bang" and the Bible calls "separation of the waters from the waters. This evolutionary period of material warming (waters) from the production of light or first radiation, to the Big-bang, was described by the Bible as the second day. And then came the third day and the fourth day which marked the beginning of the post-Big Bang period which were periods of the creation of the stars from the same material or waters according to the Bible.

Then came the fifth biblical day, the period that witnessed the appearance of living beings from microscopic underwater beings to birds.

And at the end or on the sixth biblical day, there was the appearance of mammals and of humans. (Gen 1:1-30; Gen 2:1-7).

These are exposed in a new way for astrophysicists, cosmologists, astronomers, theologians, philosophers and my other readers, the Biblical Model on which all the other Models that will support it should be built.

2.3. Original elements, original particles, original matters

Why do we usc the expressions : "original element", "original particle", or "original material" instead of elementary particle, or fundamental particle? Indeed, in science we mean by elementary particle, or fundamental particle, a particle whose composition we do not know : we do not know if it is made up of other smaller particles.

But, since elementary particles include the "fundamental fermions" (quarks, leptons, and their antiparticles, antiquarks and antileptons) that compose the material and antimaterial, as well as bosons (gauge and Higgs bosons) which are vectors of forces ,and play a mediating role in the elementary interactions among fermions (Cfr : Wikipedia; elementary particles), and not knowing with precision what was the matter before the Big-bang, from which these elements come from, we call them either the original elements, or the original particles, or still the original matters, in reference to God who is the origin of all things or the origin of origins according to the oriental philosophy. This reference to God has nothing to do with a belief just as the names of the planets of the solar system have their origins in Greek mythology.

In addition, elementary particles are the smallest physical objects of which matter and the forces of the universe are made up. They are packets of energy characterized mainly by mass, and other numbers such as electric charge. They acquire these properties only from the original particles, emanating from the dual characteristics of God. (Cfr : Presentation of the Divine Principle : ed. Culture et Paix, Paris, 2013, Pp 26-43).

Section 3 : The notion of biblical times

If the biblical scheme of creation was previously rejected by science, it was because of the notion of time.

Indeed, the biblical chronology shows a Universe that was created in six days and man appeared 6,000 years ago. The scientific chronology, on the other hand, indicates that the Universe was created 13.8 billion years ago and the oldest known fossils of the human lineage date back 7 million years. Consequently, scientists and atheists have called the Bible a book of myths. The books in the series "The Verses of the Scholars" have put an end to this definition with scientific arguments.

3.1. Biblical days are periods

As I said at the beginning, Moses taught people who knew how to distinguish periods from the moon, but who did not have the notion of time like us. So, instead of complicating their lives with bulky numbers, he preferred to teach them the history of creation in a very simple way by dividing it into six days which, in fact, means six periods or six stages. Is this my interpretation? Not at all.

The Bible being a whole, despite the biblical verse : "But one thing, beloved, which ye must not ignore, is that one day is as a thousand years before the Lord, and a thousand years are as one day" (2 Pet. 3 :8), no Christian or Jewish theologian has been able to make a connection between this verse and the days in the biblical scheme of creation, believing that the latter was the work of a word. It took a new revelation of God to the Moon couple for reaching the clarification on the basis of this verse.

Indeed, considering the books contained in the Bible as a whole by Christians, whose author, God, is like a hero in the shadows, the biblical day as quoted in the first chapter of Genesis is well defined in the same Bible in two verses :

Ps 90 : 4 *"A thousand years to God are like one day, they are like yesterday already gone, like a short hour in the night.* "... "

2 Pet 3 : 8 *"But there is one thing, beloved, which you must not ignore, that before the Lord a day is as a thousand years, and a thousand years are as a day.* "

Rev. Dr. Sun Myung Moon is the first religious man to have associated them with the biblical scheme of creation in the Creation Principle; a branch of the book of the Divine Principle.

There are two very important terms in these verses, namely "to ignore" and "like". First of all, we should not ignore the notion of time. Secondly, there is the comparative term "like" : "a day for God is like a thousand years". Here, the word "equal" has not been used, but rather "like". This simply means that a biblical day can represent as many days, as many years as a cycle or an era can take.

So to say that creation was made in six biblical days has nothing to do with our twenty-four hour days. Rather, it represents six great stages which, as science has discovered, took thousands, millions, even billions of years.

Since Moses, the writer of Genesis, later referred to the day after the six days as a weekly Sabbath, many believers and non-believers alike thought that every day of creation must have been 24 hours.

Moreover, the biblical story does not exclude the possibility that during each "day", or period of creation, some of the major creative acts occurred gradually rather than instantaneously. Hence the terms "evening and morning" used to move from one period to the next. For more details, I refer you to Volume I, the explanations which are taken from the book "Presentation of the Divine Principle". (Cf : Presentation of the Divine Principle : ed. Culture et Paix, Paris, 2013).

Finally, it should be noted that in Genesis, Moses wrote that the Universe had a beginning when material and life appeared in stages, gradually over several periods. Who was Moses? The greatest scientist that the Universe has ever had.

Ge 33:11 "*The LORD spoke with Moses face to face, as a man speaks to his friend*".

Section 4 : The Three Great Stages of the Universe or Biblical Water Phases

Throughout our book, we will have to talk about the word Big-bang (Pre- Big-bang or after Big-bang). In my previous books, I had already explained it, but a revision is always necessary for those who have no notions about this scientific word.

The Universe knows three great periods according to the Universal Cosmological Model ; the pre-Big Bang, the Big-Bang and the post-Big-Bang.

In the basic statement of the Universal Cosmological Model, the biblical water or universe has gone through three phases.

4.1. First phase : water or original particles

Ge *1 :2 The earth was formless and empty, and darkness covered the surface of the watery depths, and the Spirit (a wind) of God moving was hovering over the waters... 3 God said, let there be light. And there was there was light. 4 And God saw that the there was light good; and God separated the light from the darkness. 5 And God called the light day, and he called the darkness night. And there was evening and there was morning : and it was the first day.*

This first phase of the universe is called the Pre-Big-bang where we see that there was already a great force called the wind of God which manifested itself around a material called water by Moses and which, for us, was composed of the original particles. This pre-Big-bang phase came to an end on the first biblical day which we will explain in the following chapters.

4.2. Second phase : water or condensed Universe

Ge *1:6 And God said, Let there be a expanse between the waters, and let it divide the waters from the waters.7 And God made the expanse, and divided the waters that were under the expanse from the waters that were above the expanse. And it was so.*

The second phase was that these biblical waters, which we have described as original particles, after their extreme condensation due to cosmic deflation, ended up exploding (Big-bang), a phenomenon described in the Bible as separation (separation of the waters from the waters) and expansion by some scientists.

This is the phase called the Big-bang in science, which took place in the second period of biblical creation.

Therefore, the Big-bang represents the zero instant of the Universe and not the beginning, because the Universe was created in a period that extends to minus X number. It was the blossoming of the condensed Universe described by the Bible as "the separation of the waters from the waters".

4.3. Third phase : from the waters, the stars were formed to the living.

Gen 1 :9 And God said, Let the water under the heaven be gathered together into one place, and let the dry land appear. And it was so.

Gen 1 :14 And God said, let there be lights in the expanse of the heaven to separate the day from the night. They will serve as signs for seasons[c] and for days and years.

Gen 1 :20 " God said, "Let the waters bring forth living creatures in abundance".

Ge 2 :6 But mist would come up from the earth and water all the ground. 7 And the LORD God formed man out of the dust of the earth,

The third phase of these biblical "waters" was that of the

creation of the stars to living beings. By this, we understand directly that it was not the waters we consume, but a material that had its origin in the Pre-Big-bang, crossing the Big-bang until the formation of the stars and living beings. This last phase is what we call the post-Big-bang phase.

We wanted to explain this word of Big-bang which represents a condensed Universe and which has grown in volume through expansion, defined in the Bible as "expansive", for those who do not yet have this preliminary notion of cosmology. We can schematize what we have just said in the following way :

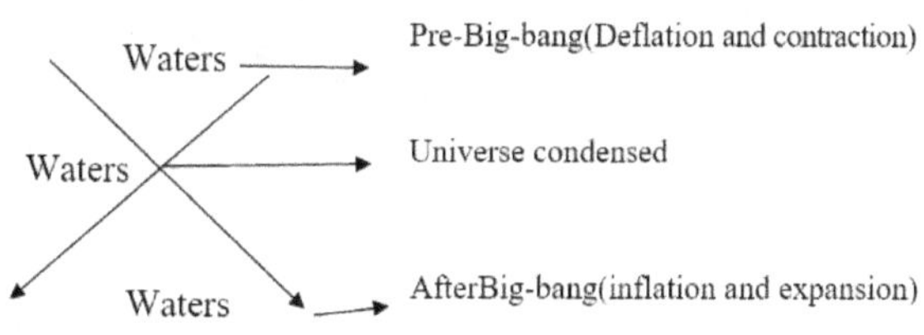

Chapter II :
The Cosmos

We still want to come back to the different notions of the cosmos because my books are read by thinkers who don't often have the notions of cosmology. The cosmos is not included in the pre-Big-Bang or post-Big-Bang, but it contains them. So, even if one day we will manage to analyses all the periods from Pre-Big- bang to after Big-bang, there will always remain an enigma, it is the Spiritual World which is also part of the Cosmos.

Section 1 : The Creation of the Cosmos

1.1. God is the creator of the Universe

"In the beginning God created the universe (the heavens and the earth). (Gen 1:1) HEAVENS : non-quantifiable spiritual realm

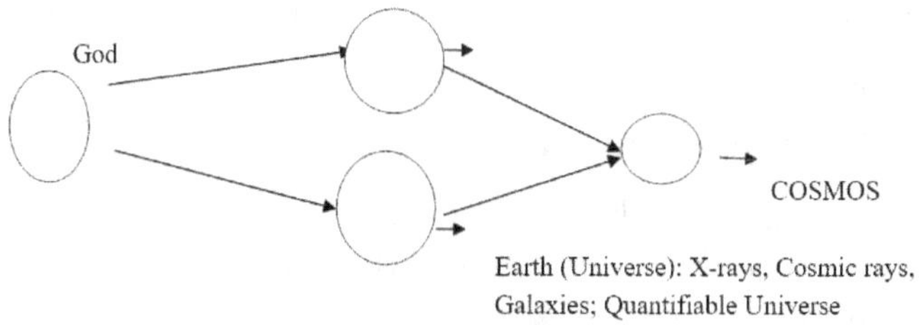

Ge 1 : 2 *The earth was formless and empty :* something that exists but cannot be felt; therefore, without form, i.e. the Universe Pre-Big-bang.

As we have said, the cosmos, according to the Universal Cosmological Model, which is in line with Rev. Dr. Sun Myung Moon's Principle of Creation, represents the invisible (spiritual) world called the visible (Universe), both worlds are called "the earth" in the Bible. Thus, when speaking of the creation of Heaven and Earth, the latter meaning the Universe, we see that the invisible world (or angelic world, or spirit world) preceded the creation of the Universe represented by the earth in the Bible.

This definition of the cosmos is different from encyclopedias which do not take into consideration the spiritual sphere. In the Bible, these two terms, the spiritual and the physical world, are represented by the words heaven and earth.

For science, the cosmos represents the Universe in all its extent, i.e. the visible part and what our instruments of observation do not yet allow us to see. But when the Principle of Creation speaks of the cosmos, it includes the spiritual world. It is this conception that is taken into story in the Universal Cosmological Model.

Thus, in the biblical verse "In the beginning God created the heavens and the earth" (Gen 1:1), the words earth and sky are in lower case because it has nothing to do with our geographical Heaven and Earth.

1.2. From the spiritual to the physical state

What materials did God create the Universe from? This was done by the transformation of spiritual energies into physical energy, called, in the esoteric sciences, "the densification of energy". This is a point that intrigues all researchers, namely how the material physically moved from the invisible to the visible, biblically described by the phrase "In the beginning God created the heavens and the earth" (Gen 1 :1).

The Bible speaks of an initially empty Universe (Gen 1:2), and later, in the same verse, we are told of the presence of something material, referred to as "waters". What happened to bring us to these "waters « ?

Biblically, there was at first a vacuum, and in the vacuum, according to science, there is energy. Where did it come from? According to Rev. Dr. Sun Myung Moon, everything comes from God.

Indeed, according to the Creation Principle of Rev. Dr. Sun Myung Moon, in God there is the Universal Prime Energy. By projecting and densifying itself in a land, this part of the spiritual world has become the material world while being visibly formless and empty. Hence the biblical verses :

Gen 1:1 In the beginning God created the heavens and the earth.2 The earth was formless and empty.

We have already explained that something that is formless and empty simply means that it is materially non-existent. Rather, it was the beginning of an empty but energetic universe.

The word densification will be the subject of many publications in the future, as well as other lines of research in the sphere from the spiritual to the physical. This is why the "Universal Cosmological Model" opens a revolution in cosmological science.

1.2.1. The biblical heavens

Gen 1:1 "In the beginning God created the heavens and the earth".

The skies we are talking about here have nothing to do with the geographical Blue Sky we see above our heads. It is rather the spiritual world which is the sphere of the spirits. It is in these spheres that we find what we call, here on Earth, the kingdom of heaven, paradise and hell. It is therefore a spiritual sphere and invisible to us humans.

So, said, what is important for scholars was to understand what the word heaven means in the verse that says "*God created the heavens and the earth*". As we have just said, this is the spiritual world and not our geographical Heaven. Hence the prayer : "Our Father in heaven" (Mt 6:9).

1.2.2. The biblical earth

Gn1 : *1 "In the beginning God created the heavens and the earth "2 The earth was formless and empty.*

As with the heavens, the earth referred to in Gen 1.1 (God created the heavens and the earth) has nothing to do with our geographical Earth. Moreover, it is also the use of the word earth and its misunderstanding that has caused the biblical scheme of creation to be trashed by scholars. And, no theologian is there to explain it to them, simply because I am the first to explain it.

The word earth here refers to the physical world known as the Universe. Did I invent this explanation to conform to science ? Not at all, because Moses was clear and no one was able to discover it. In the next verse, Moses says : "The earth was formless and empty" (Gen 1 :2). Something that is formless and empty is something that is immaterial or materially non-existent. Even visible smoke has a form, but here Moses is clear : in the beginning, the Universe he called "the earth" was formless and empty. The present Earth with these structures had only appeared in the third stage, after the Big-bang described by Moses as the third day.

With these clarifications, now the biblical story of creation, which sheds light on how the Universe began, i.e. in the vacuum, enters science. We have called this great restitution the "Universal Cosmological Model".

The model according to which the Universe began in a vacuum was developed in a scientific way by the Italian scientist Gabriele Veneziano without him knowing that he had joined the biblical model that has become the Universal Cosmological Model to this day. If he had known this, he would also have had to discover that the fossil radiation is a remnant from before the Big-bang through the verse "God said that light be and there was light ! "This occurred as the pre- Big-bang material began to condense, while warming up during the phase of cosmic deflations.

It is this last phenomenon, described by the Bible,

has been forgotten in the Model developed by Gabriele Veneziano. This is why we say that the Universal Cosmological Model swallows up and corrects the whole Cosmological Model. Several other models have speculated about the Pre-Big-bang, but I am the only one who speaks about it with conviction.

Moreover, I was surprised, at the end of my work on Volume I, to discover that my research on the updated restitution of the biblical model has more or less reached the string model. As I said, string theory was created by the Italian physicist GabrieleVeneziano in 1968.

Here is an excerpt from his speech :

"*In my scenario, we start from an almost empty, very, very diluted Universe with very weak interactions between the components. In this near-emptiness, however, there are instabilities that can become catastrophic and lead to gravitational collapses like a black hole. And the Universe as we know it would come out of this initial near vacuum through this black hole*" (Cf : Gabriele Veneziano, interview by Guy DUPLAT)

(Cf. :
http://www.lalibre.be/archive/qu-y-avait-il-avant-le-big-bang ; published on 15/12/2005).

The Divine Principle of the Moon couple defines the Universe as being the visible aspect of the invisible God.

The cosmos includes the spiritual world (invisible) and the physical world (Universe). The Universe contains, in its empty spaces, galaxies, dark material, dark energy and other wandering bodies. The galaxies contain black holes, stars, planets, exoplanets and other bodies roaming in the vacuum such as asteroids, comets and meteorites. Planets contain continents, gas and water in various forms. The continents of the earth contain the countries.

1.3. Cosmos, sphere outside the universe

As we said in the introduction to this chapter, the cosmos represents the spiritual world and the physical world, translated by the phrase "God created the heavens and the earth".

When we speak of the pre-Big Bang and post-Big Bang, we are in the physical sphere. The Cosmos encompasses this sphere. Therefore, we cannot classify the cosmos in the pre-Big Bang sphere. The pre-Big-Bang sphere has a beginning, hence the biblical phrase 'God created'. The creation of the universe began when the word earth was first mentioned in the Bible and described as "a formless and empty earth».

In Volume I we said that at the very beginning of the Universe there was only a vacuum as the Bible says "the earth (Universe) was formless and empty". But what tells us that we were only facing two spheres where we could distinguish between the spiritual sphere (the heavens) and the material sphere (the Universe), especially since they were, for a fictitious observer, visibly empty? And how is this emptiness filled today with physical structures?

The answers come to us from science and mainly through quantum physics. Speaking of a formless and empty earth, therefore materially non-existent, physicists have found that the vacuum is filled with physical energies that can be virtual or elementary. This is the answer to the first question that allowed us to distinguish between heaven and earth.

In the heavens we find particles that cannot be physically detected. However, through divine radiation or projection, a sphere was created where these particles became dense in relation to the spiritual elements, becoming what are called virtual particles, contained in a formless and empty earth or the Universe at its beginning.

The second question was : How is this emptiness filled today with physical structures? This question can also be formulated in the following way : "How can we imagine the beginning of the Universe? As I said, the answer comes to us from science in quantum mechanics, but we preferred the development of the physicist and chemist **Ilya Prigogine** who wrote in "Time to Become" :

"An essential aspect is the passage from virtual particles to real particles. Virtual particles are particles that are created in pairs, live for a certain amount of time and then recombine as a result of Heisenberg's uncertainty principle. Real particles, on the other hand, no longer have to recombine; they are no longer held together by a thread. So the creation of the Universe, the creation of material as we know it, corresponds above all to a creation of possibilities, creating both disordered and highly organized phenomena.»

(Cfr : http://www.matierevolution.fr/spip.php?article38Le vide,... not so empty, le23/06/2010, by Robert Paris).

The vacuum is therefore full of virtual particles. We then understand that reality is nothing other than a transformation of virtual particles.

Section 2 : Differences between Biblical and Geographic Heaven and Earth

2.1. Biblical heaven and earth

Ge 1:1-2 *"In the beginning God created the heavens and the earth. The earth was formless and empty.»*

We have just explained, in the previous section, that the heavens spoken of in this verse allude to the Spiritual World which is invisible to us. The earth, on the other hand, through its definition of "a formless and empty earth", alludes to the physical world called the Universe. Here we are talking about its beginning, that is, the pre-Big Bang Universe began in a vacuum.

2.2. Geographic Sky and Earth

2.2.1. The Geographical Sky

Talking about geographical sky and earth is like talking about the beginning of space-time. All astrophysicists say that space and time in the Universe started from the Big-bang, joining the Bible. How did this happen?

Gen 1 :*2 "...a wind of God moved upon the waters"...*

Ge 1:6 *And God said, "Let there be an expanse between the waters, and let it divide the waters from the waters. 7 And God made the expanse, and divided the waters that were under the expanse from the waters that were above the expanse. And it was so. 8 And God called the expanse heaven. And there was evening and there was morning : and it was the second day....*

In verse 2 above, we are told : "a wind of God (energy)", which is an attribute of God, was moving in the water which was the representation of the original particles. Starting from Genesis verse 6, God separated this water; this is the Big-bang : "Let there be a expanse between the waters" and "God called the expanse Heaven".

From now on, we have a clear definition of the geographical Heaven. It is the expanse or empty space that separates everything and encompasses everything. This empty space begins even where I set my feet. This empty space we neither know its beginning nor its end. This space that encompasses the entire observable and unobservable Universe contained in the cosmos. We can say that geographical space is the visible part of the sky. Therefore, we can also call it the sky, but it is only content.

Thus, it is from this verse : "God separated the waters from the waters", forming the expanse, that space and time were created in our universe according to Big-bang. However, before the Big-bang, we do not know how to define the times that the required biblical events described on the first day (Gen 1:2-5). But because there was a wind from God was hovering over the original material, the warming that created the emergence of light, and the extreme temperature of the condensed Universe, we can say that there was also time in the front. And, since we have spoken of the abyss, which means deformation of the fabric of space, we can say that there was also space.

With these clarifications, what can we say ? Is the biblical scheme of creation complete or incomplete ?

2.2.2. The geographical Earth or planet Earth

We said that the earth, which is spoken of in Genesis 1:1 where it is written "In the beginning God created the heavens and the earth", had nothing to do with planet Earth. In this verse it was about the Universe at its beginning.

The Bible began to speak about planet Earth from the third period of creation through the following verses :

Gen 1:9*God says, "Let the waters under the heaven be gathered together in one place, and let the dry land appear". "And so it happened.10 God called the dry land earth, and the mass of waters the seas. God saw that it was good.*

As we can see, it was at this point that the Bible spoke of our planet Earth or geographical Earth and that this name had nothing to do with the verse "God created the heavens and the earth" which represented the spiritual and physical spheres.

2.2.3. Summary

The great sets of the universe :

1) Cosmos

Ge 1 :1 In the beginning God created the heavens and the earth

– Spiritual World (Heaven)
– Physical world (earth)

Spiritual world : heaven

– Kingdom of Heave
– Paradise
– Hell

Physical world

Ge 1 :2 A formless and empty earth

– Universe

2) Universe

Space, Dark material, Dark energy, Galaxies

3) Galaxies
- Stars
- Planets
- Exoplanets
- Other celestial objects

4) Planets and Exoplanets
- Continents
- Waters
- Gas

Chapter III :
THE Pre-BIG BANG

For science, before the invention of the Universal Cosmological Model, the Pre-Big-bang was a non-speculative period, as a Frenchman answered me in a blog. The latter said to me :

"This being the case, no one knows how to model precisely this mysterious zone below Planck's length and time, it remains an aporia".

(Cf. : zevengeur.wordpress.com/2014/01/16/histoire-de-l'univers Part 6 : Janus model,
the future of cosmology ?).

But, in the light of all that we are demonstrating in this book in the framework of the Universal Cosmological Model, it will no longer be so and scientists will get back to work.

Section 1 : Generalities about the Pre-Big-bang

1.1. The study of the Pre-Big-bang is physical

We have defined the Universal Cosmological Model as the scientific restitution of the biblical scheme of creation. This pattern is contained in the Bible in the book of Genesis, chapters I and II. The author of this scheme, which we call the Biblical Model or the Biblical story of creation, is Moses. I am the first in the Universe to offer an explanation of this scheme which until now nobody has understood even the first verse which says that "God created the heavens and the earth".

In chapter I of Genesis, Moses makes God the sole creator of the Universe. In Chapter II, on the other hand, in addition to attributing the creation of the Universe to God, Moses does indeed enter into cosmology with evolutionary phenomena, from creation, by evoking the climatic development of the Universe, to the way in which we have gone from asexual to sexual, using man as a case in point.

All this is to say that we do not come, within the framework of the Universal Cosmological Model, to impose the Biblical Model as an act of faith. If Fred Hoyle's team left us the nucleosynthesis to explain the evolution of material from atoms, the Universal Cosmological Model offers us rather the properties contained in the virtual particles that allowed their assembly or fusion until they became the real particles.

We cannot go beyond these particles given our current knowledge. However, in Volume I, I presented the path that went from the spiritual to the physical by the creator of the Universal Cosmological Model. Not even the Bible has touched on this, although the book of the Unification Thought, another teaching manual of the Rev. Sun Myung Moon, has brushed against it with a religious language that I cannot include in this manual. Even if I went further, what is important for physics and cannot go unnoticed is the beginning of the Universe in the vacuum.

But how did these energies of the vacuum come into action? Science calls this attraction at the microscopic level "the graviton". But what is the causal power of this graviton? The answer is rather to be found in the assiduous study of the Principle of Creation and the thought of the unification of Rev. Sun Myung Moon, for this was done on the basis of the dual characteristics of God endowed to each element from virtual particles. If we can scientifically rectify this sentence, we can say that it is from the dual characteristics that these particles have.

So, if the Standard Model has brought us nucleosynthesis, the Universal Cosmological Model has just brought us the principle of creation which explains the origins of force at the microscopic level and its manifestation as studied in quantum mechanics, whose synergy we have named "the physics of creation".

The universe according to the Bible began with a vacuum described as a formless and empty earth, that is, nothingness and chaos. Astrophysicists have discovered that the vacuum is full of energy. It is these energies, by entering into the action of giving and receiving, as described in the Principle of Creation and studied in quantum mechanics, that led to the real energies. These phenomena have not escaped the Bible and this has been described by the biblical verse in the Jerusalem version "a wind of God was moving the waters" (Gen 1:2).

In the Scientific Documentary : "THE UNIVERS, HAS IT NEVER STARTED ? "a lecture by the French astrophysicist

ETIENNE KLEIN, the scientist is wary of the singularity of the Big-bang, which he considers to be a transitory stage, and demonstrates that the study of the pre-Big-bang cannot be considered metaphysical, but rather physical.

As soon as the point about the Pre-Big bang is raised, astrophysicists stop breathing. They see this point as if one were asking a believer "who created God» ?

All the answers proposed so far concerning the Pre-Big-bang have only disappointed everyone. No one thought the answer was already in the Bible.

Indeed, the story of creation according to the Bible has already been thrown away by scientists. It had to be reconstructed in the "Universal Cosmological Model" so that the scientists discovered that they had so many answers in their hands.

Thus, with the Universal Cosmological Model, revealing the mass contents of the Universe before the Big Bang, scientists will once again turn back time, by studying more closely the fossil radiation that would be the first radiation that appeared before the Big Bang, described as "light" by Moses. But, before that, let's look at the basic principle that was at the origin of the creation of the universe.

1.2 Knowledge of the pre-Big-bang period in ancient times

In Volume I, we have spoken in details about the knowledge of the pre-Big- bang by ancient philosophers, particularly Empedocles.

We know that the pre-Big-bang is a period which is not yet scientifically recognized. It is in the biblical scheme of creation, written 1,500 years BC by Moses. A similar scheme was described by the philosopher Empedocles who lived in the 5th century BC without having had biblical knowledge.

While for science time and space did not exist during this period, through the Bible we discover that it was a very eventful period, described as the first day, as we can read :

Ge 1 :2-5 *"The earth was formless and empty, and darkness was over the surface of the deep, and the Spirit of God was moving over the waters. 3 And God said, Let there be light. And there was light. 4 And God saw that the there was light good; and God separated the light from the darkness. 5 And God called the light day, and the darkness he called night. And there was evening and there was morning, and that was the first day.*

As for Empedocles, he described the events that took place Pre-Big-bang as follows : "Originally, all the elements, united by Love, formed a homogeneous mass called the "spherus". (Cf. : Empédocle, Wikipedia). Here, we are talking about the elements that were united, before we arrived at a homogenous mass called the "spherus", effectively demonstrating that there was a force that he called "love" and that it was also an eventful period. We have developed the contents of all these periods from before the Big-bang to the Big-bang, whether they are described in the Bible or in the ancient models in Volume I.

With these two testimonies of Moses and Empedocles, two great scholars who did not know each other and never read about each other, it now becomes dishonest for an astrophysicist, whether American, French, Russian or Chinese, to answer to someone who does not know what was the Pre-Big-bang. The only answer now, while waiting for science to pronounce itself :

"The events before the Big-bang are described in the first cosmological book, namely the Bible, on the first day of creation, in the book of Genesis. But all scientists are currently

working on this to shed light on the events that took place during this period".

This answer corresponds to the injunction that God gave me, namely : "*Tell everyone to be inspired by the biblical story of creation*". God did not tell people to believe, but to be inspired; that is, to study it and deepen it so that each person can come to the true conclusion himself; just as he left me to fend for myself.

In addition, to start answering many questions based on the "why", it becomes imperative for the leaders of the Unification Church, who are called to end all denominations they create after the ascension of Rev. Dr. Sun Myung Moon, to start organizing seminars based on the Principles of Creation and the Unification Thought for scholars.

As we did in Volume II, where we devoted a chapter to respond to atheist physicists (the universe of atheists), we will also devote a whole chapter here, at the end, to answer the various questions that some Internet users have asked me in different forums on the web, mainly in Quora.com, concerning especially the pre-Big-bang and the Big-bang.

Section 2 : The physics of creation

The Physics of Creation, which we discussed in Volumes I and II, includes Rev. Dr. Sun Myung Moon's Principle of Creation and Quantum Mechanics. The Principle of Creation brings out the intrinsic properties of particles that give them the ability to move and through quantum physics, astrophysicists present us the results of this movement that led to material from live particles in material.

We cannot talk about what happened in the pre-Big-bang era without offering our contribution by highlighting the properties of virtual particles. We have already done so in Volume I, so we will draw several paragraphs for the teaching of astrophysicists who will read only this Volume before deciding to return to Volume I and II for more details.

The creation story is attributed to Moses and is contained in the Bible. It is thanks to other notions that we were able to join the puzzles together to create the Universal Cosmological Model.

2.1. The contents of the physics of creation

The basic teaching that has enabled us to understand the origin and functioning principle of the Universe is The Principle of Creation of Rev. Dr. Sun Myung Moon. Through this principle, we understood that no one can better explain the Universal Cosmological Model without the elementary knowledge of the Word of God, cosmology and the Principle of Creation contained in the Moon couple's book of the Divine Principle. Added with the notions of quantum mechanics, this forms the physics of creation.

Therefore, when speaking of the Physics of Creation and the Universal Cosmological Model, one departs from any judgment that one may bring to these sacred writings which are the Divine Principle, the Bible or other sacred books.

I myself bear the responsibility for the judgments of the selected writings or verses. So, I offer you these two definitions :

– The Universal Cosmological Model is the scientific restitution of the story of the biblical creation. Its scope begins when the biblical creation narration speaks about « the shapeless and empty earth" which shows the beginning of the universe originally till the last stage when God created one man and one woman.

– The Physics of Creation encompasses all the branches of physics or all the thoughts of physicists, religious or philosophers who have developped their thoughts about the world creation from emptiness to the appearance of the first particles.

The book "THE VERSETS OF THE Scholars" as for itself, extends its field of action until the second advent of Christ which will be developed in the other Tomes. (F. PHANZU; the cosmological according to biblical and ancients models, Éd. Librinova, Paris 2021, P.97).

We have said that all the scientific theories developed today, as having led to the formation of primitive particles, are phenomena which took place in the time of chaos Pre-Big-bang according to the Universal Cosmological Model and are now part of creation physics.

Indeed, the Big-bang being characterised by an exponential acceleration of the Universe, called "inflation", it would have been impossible during this time span estimated in milliseconds, in this cosmic plasma, for elements to be created from virtual particles to atoms in the space of 10-43(10 exponents -43) seconds to 3 minutes. It was rather a time of liberation of what has already been created.

For astrophysicists, during this exponential acceleration, the Universe expanded strongly and this allowed it to cool down so that the subatomic particles could fuse together. Beginning at 1015 ℃, what was the temperature and volume of the Universe so that at three minutes, the particles would merge ? It's up to the scientists themselves to study and rectify this argument, which was theoretically put forward in the Standard Model.

Here is an illustration of the Big-bang or the separation of water from water

Image :

https://qph.fs.quoracdn.net/mainqimg4d67c1855b85cf03

*"With such violence during the Big Bang event, did the elements have time to merge in the space of a thousandth of a second to 3 seconds; starting from virtual particles, some cohesion particles - antiparticles through the strong interaction of quarks up to form the atom"? (*Flavien Phanzu Mwaka).

What science has not been able to restore so far is that if the model of cosmological inflation has offered a solution to the problem of the horizon as well as to the problem of platitude (Cfr : Wikipedia; Cosmological principletists), the scientists rather create some terms about the creation of the primitive elements starting from the virtual particles to the atoms. But this model has been presented until then as the one which also solved the problem of the constitution of atoms. If this is the case, scientists should also accept the biblical sentence saying that "God commanded and the thing was done" because even if you are a magician between "commanded and done" there is always a gap of seconds. (F. PHANZU; the cosmological according to biblical and ancients models, Éd. Librinova, Paris 2021, P.141).

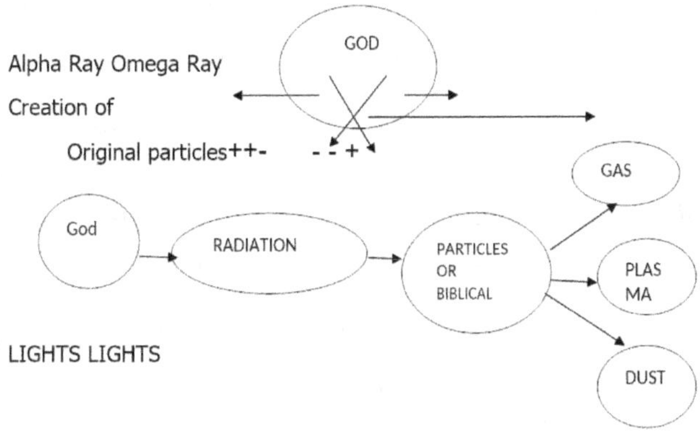

Rev 22 :13 "*I am the Alpha and the Omega, the first and the last, the beginning and the end*".

This scheme can be understood through the following scientific sentence :

"Since the origins of the history of the Universe, the latter was created of a field of very high energy radiations which during a" hurly and burly" has generated material particles to the tune of about 1% from the initial energy. The latter is still present today with a very low energy density in the form of the famous residual cosmological background". (Cfr:zevengeur.files.wordpress.com/2018/07/instant-zéro).

All these reasoning of scientists, according to them concern the post-Big- bang period; but, with regards to the Universal Cosmological Model, the creation of the primitive particles took place during the before Big-bang period, even if after confinement, it firstly behaved like a plasma soup.

God is the source of energies. So, after having actuated the circular movement of the Universe, what Moses could have said by "God says that the earth rotates" ; everything began to be created from there. God is action. This truth, even science cannot contradict it. We can show it by this statement :

"The vacuum in quantum mechanics is not really empty. All the time, everywhere at each moment around us, pairs of particles and antiparticles are created out of nothing". (Aurelien Barrau, astrophysicist, interviewed by futura-Sciences, By Jonathan SARE, Futura-Sciences 08/01/2016). So let's look at the notions that can help us understand how the breaking of the symmetries of virtual particles in an empty universe has been carried out so that we end up with real particles. However, my introduction will allow scientists to enrich it. As I said the Universal Cosmological Model does not come to close, but to allow astrophysicists to enrich it, especially the pre-Big-bang, which can automatically correct several post-Big-bang theories contained in the Standard Model and other models.

2.1.1. The principles of creation

The Principle of Creation, a branch of the book "Divine Principle" by Rev. Dr. S.M. MOON and the series of books "The Verses of the Scholars", form the basis for scientists to become part of the scientific understanding of God, as well as the origins of the universe up to man.

In order to know the universe and its principles better and to be able to explain it better to scientists without using parables and symbolism, but an appropriate scientific language, Rev. Dr. Sun Myung Moon, whose life we have traced in Volume I, was obliged to study engineering in Japan. These conclusions, which better explain God, the origin of the Universe and its eternity, are recorded in the Principle of Creation, a chapter of the Divine Principle, as well as in his many works.

The originality of the Principle of Creation is that it integrates the causal aspect and seeks to explain the functioning of the Universe based on the characteristics of God himself.

From this Principle of Creation, we have sorted out only four points that are of interest to the physical science and which will be part of the "Physics of Creation", just like quantum physics. Thanks to these four features of the Principle of Creation, several riddles of the universe will be solved by scientists, as Ms. Moon said in one of these recent speeches :

"If you take into consideration what is taught in the Divine Principle and start to apply it in the different areas, you are going to discover more, and you will be able to explain a lot of things (enigma)."

2.1.1.1. The four features of the Principle of creation

The four features of the Principle of creation ; branch of the Divine Principle book of Rev. Sun Myung Moon who will be part of the physics of creation and will now be the subjects of study are as follows :

– The dual characteristics of God, which explains how the universe functions ;

– The Universal First Energy : which describes the source of energy and its extension in the universe ;

– The period of growth of the created beings, which reveals that the biblical days are the periods ;

– The spiritual and physical world who's the Rev. Dr. Sun Myung was the only one to describe the relations that unit both and form the cosmos.

2.1.1.1. The dual characteristics of God

The universe is dynamic. The dynamics of the movement of the Universe is observed by scientists through the writings below, which they do not manage to give a convincing explanation :

" *Without claiming that our Universe turns round, we have to admit that everywhere, from the closest to the most distant objects, from the most modest to the most gigantic, we observe a great unbridled waltz : proper rotation movements of planets, stars, galaxies, revolution movements of these same planets, double stars...* "

(Cf. : H. GIÉ : Le moment cinétique à travers l'Univers * BULLETIN DE L'UNION DES
PHYSICIENS P.427 N ° 712, Paris).

This dynamite finds its essence in the dual characteristics of God endowed to these particles. Some scientists, who try to explain this force, begin to define it as gravitons. This is why the knowledge of the dual characteristics of God, explained in the Principle of Creation, is very important for a better understanding of the dynamism within the Universe.

As we have said, the universe is a divine projection. Here, the word projection must be taken in the sense of giving birth. God gave birth to the Universe and the Universe can only reflect these characters. We are not in religion but in physics because we are going to base ourselves on these physical characteristics, which make everything in the Universe dynamic.

Indeed, the dual characteristics of God are already beginning to manifest themselves through virtual, elementary and subatomic particles. Dynamics at the level of elementary particles is known as spin, which is their fundamental intrinsic characteristic. Therefore, the effects of the dual characteristics of God already begin to manifest themselves from the elementary particles. For example, we notice that an electron only returns to its initial state by a double spin : this is called a spin particle ½.

The universe being the visible extension of the invisible God, we have said, we can only better understand its functioning through the common and visible characteristics of creation as explained in the Principle of Creation. These characteristics have their source in the invisible God manifested through creation, as confirmed in the following verse :

Rom 1:20 "*For the invisible perfections of God, his eternal power and divinity, are seen as to the naked eye, from the creation of the world, when they are considered in his works. They are therefore inexcusable*".

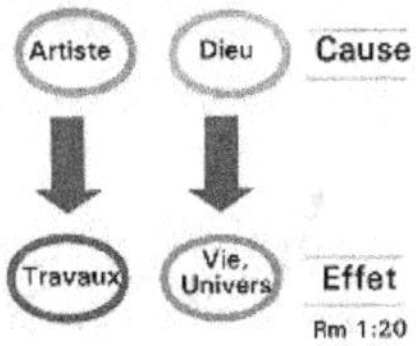

Source : The Divine Principles, Level 4, p.13

Saint Paul's great wisdom in this verse (Rom 1:20) reminds us that anyone who says he does not know God is not analyzing the creation and himself well. The Rev. Dr. Sun Myung Moon was the first to point it out in the Principle of Creation as the "Dual Characteristics of God". During his researches, he came to the conclusion that all beings are made up of the following dual characteristics :

1) All materials, from particles to humans are made up of an invisible entity and an external form called mass by the scientists according to the following table :

CREATION	Invisibleentity (internal)	Mass [external form]
Male	Heart [soul] and Spirit	Body
Animals	Animal spirit	Body
Plants	Plant spirit	Body
Molecules	Inherent nature	Energy/material
Atoms	Inherent nature	Energy/material
Particules	Inherent nature	Energy/material

As Saint Paul has pointed it out, by examining all these characteristics that are universally present in the various elements of creation, we come to understand now the nature of God : He is the First Cause of the whole universe as well as the partner-subject himself who has the harmonious dual characteristics of inner nature and original outer form.

Everything that is scientifically unexplained in his creatures finds its causation in their invisible entities. The Bible describes this by the following verse "Then the Lord God formed a man[c] from the dust of the ground and breathed into his nostrils the breath of life, and the man became a living being" [Ge 2 : 7]. The dust of the earth representing the mass, and, the invisible entity is represented by the breath of life. A detailed explanation is contained in the book "Presentation of the divine principle : ed. Culture and Peace, Paris, 2013, Pp 21-25 ".

2) All materials, from particles to humans, are made up of a positive entity and a negative entity according to the following tableau :

CREATION	Positive entity	Negative entity
Man	Male	Female
Animals	Male	Female
Plants	Etamine	Pistil
Molecules	Cation	Anion
Atoms	Positive	negative
Particles	Positive	negative

Having thus discovered these invisible dual characteristics in creation which in fact is only the effect, Rev. Dr. Sun Myung Moon infers that the origin should come from the cause, that is, from the Creator who is God. Consequently, the Universe is governed by laws and principles coming from God himself who is the incarnation of these same dual characteristics : invisible-mass, positivity- negative, masculinity-femininity. We must insist on the fact that the invisibility of God and his external form do not exist as independent entities, but in harmonious reciprocal relationship.

According to the principle of creation, the branch of the divine principle of Rev. Dr. Sun Myung Moon, we read this :

"In the nature, we find that within all creation there is a reciprocal relationship between positivity and negativity. For example, atoms are formed from the reciprocal relationship between positive elements and negative elements. The atoms themselves, composed of the particles, have positive or negative characteristics. Based on such characteristics, two or more atoms maintain a reciprocal relationship, forming molecules. All plants and animals exist and are renewed by virtue of myriads of interrelated inner and outer relationships

of positive and negative elements. ". Humanity is made up of men and women and human society exists, multiplies and develops through reciprocal relationships between men and women."(Cfr : Divine principle, level 4, Pp.15-17)

While mixing these tableaus, the Rev. Dr. Sun Myung Moon came to the conclusion that the relationship between positivity and negativity of beings is similar to that which exists between the invisible entity of each body and its mass.

Thus, the creation is made in such a way that everything exists by a reciprocal relation of positivity and negativity. From that several laws flow that science has formulated as Coulomb's law which says that "two charges of opposing signs attract" and tend to return to unity, I added. By this, I also conclude that gravity is one of the consequences of these characteristics.

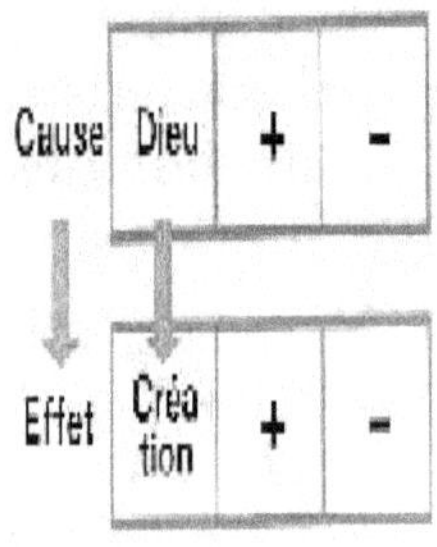

Source : The Divine Principles, Level 4, p.17

Through all this, we understand that the rotation that celestial things animate finds its essence from the spins of the particles. But what creates these dynamisms called spin from elementary particles ? It is the projection of the divine characteristics found in each particle.

In fact, all the characteristics masculinity-femininity, positivity-negativity, invisibility and mass being united and in harmony in God are called original dual characteristics. Being projected into creation, they tend, according to the Principle of Creation, to return to unity, while creating multiplication and development.

The Bible describes this in the following verses : "God created man in his own image, in his own image he created man and woman" (Gen 1 :27). Man and woman have an invisible spirit within them and an outward form (mass). In addition, man represents positivity and woman represents negativity. The source of all this is God according to the above verse. All these characteristics will also result in the union of man and woman, and Moses brought this out as a law in a figurative sense

In fact, all the characteristics masculinity-femininity, positivity-negativity, invisibility and mass are being united and in harmony in God are called original dual characteristics. Being projected into creation, they tend, according to the Principle of Creation, to return to unity, while creating multiplication and development.

The Bible describes this in the following verses : "God created man in his own image, in his own image he created man and woman" (Gen 1 :27). Man and woman have an invisible spirit within them and an outward form (mass). In addition, man represents positivity and woman represents negativity. The source of all this is God according to the above verse. All these characteristics will also result in the union of man and woman, and Moses brought this out as a law in a figurative sense : " *This is why a man leaves his father and mother and bonds with his wife, and they shall become one flesh*" (Gen. 2 :24).

This is the most fundamental law of the Universe. For those who have read Volume I, it is on the basis of this law that Empedocles managed to explain the functioning and dynamism of the Universe in the 5th century BC.

Therefore, a particle wandering in the Universe is a particle that is searching for its soul mate, called an antiparticle.

A single flesh means the re-creation of the original image, i.e. the image of God. It is this law, we have said, that governs the whole Universe, making it dynamic through rotational and gravitational movements. Unity and unification encompass the physical term as fusion. This can be explained by the diagram below.

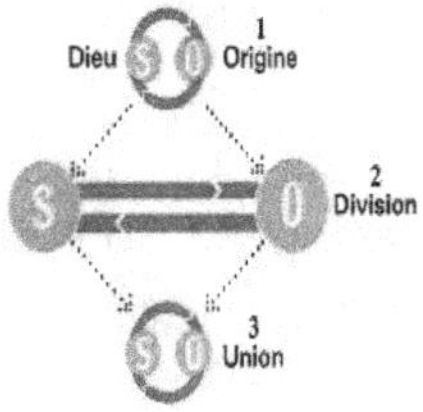

Source : The Divine Principles, Level 4, p.24

The above diagram, referred to as the foundation of the four positions in Rev. Dr. Sun Myung Moon's Principle of Creation, is not only the greatest formula, but also the greatest solution for understanding all the mysteries of the Universe. It is characterised by the action of origin, division and union.

In this scheme, God as the origin of All has in him, in his invisibility, all the dual characteristics represented by S, i.e. Subject as active character and O,

i.e. Object as passive character, but both are in harmony. In creation, the projection of the Subject and the Object is divided, starting with the particles-anti- particles, however virtual they may be. According to Coulomb's law where two charges of opposite signs attract each other, the subject (particle) and the object (anti-particle) will interact and annihilate each other. According to the Universal Cosmological Model, a new particle combining the two properties will be created.

Given the collision, the result will be that the new one will have a dominant plus over a minus or vice versa. This will lead to more new combinations or interactions until the elements are created, or what astrophysicists call the domination of material over antimaterial, an enigma which they cannot explain.

The study of God's dual characteristics in creation brings to light what we call "the uniformity of creation". It is on the basis of the functioning of this uniformity that all the riddles of the Universe can be solved. Thus, it is from the study of the dual characteristics of God that the "theory of everything" can be born, or perhaps, I myself do not know if I gave birth to it, if my explanations respond to the concerns of scientists.

As I said, the dual characteristics of God are part of the Creation Principle, a branch of the Divine Principle of Rev. Dr. Sun Myung Moon, part of the doctrinal teaching of his church and which I cannot teach in this manual. We have drawn, as we have said, the parts that are only theoretical physics.

(Cf. : Présentation du Principe divin : ed. Culture et Paix, Paris, 2013,).

2.1.1.1.2. Positivity (+) and negativity (-) within entities

strength of the 4 elem
strength of the 4 elem

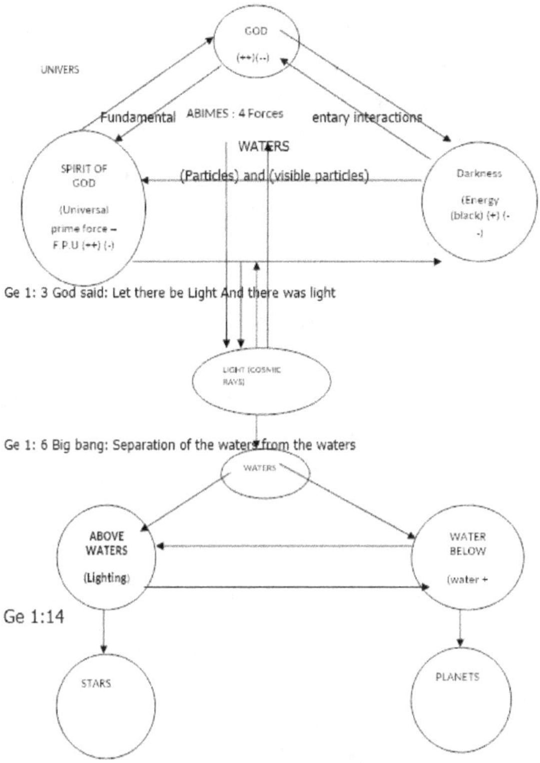

In the diagram above, we see that God has, in him, according to the Principle of creation of Rev. Dr. Sun Myung Moon to which I align myself, the two inner traits of positivity and negativity in such harmony that they cannot be separated. His whole creation inherited his characters. Positivity is found in the pro-tone; and negativity, in the electron. However, in these two particles called nucleon, there are, despite their dominant charge, the two characteristics, positivity and negativity, which are at the origin of their continuous rotation, contributing to the dynamics of the universe. This rotation, started at the particle level, is called "spin".

We know that the atom is made up of smaller particles : it is made up of electrons (negatively charged particles) ; and a nucleus made up of protons (positively charged particles) and neutrons (without electric charge).

These united particles form the atom, the smallest part of a body, and transmit this divine dynamic to it. The atom in turn transmits it to the matter which constitutes the universe. This causes all bodies, including the stars, to be composed by the characteristics, which make them rotate around their axis, forming the circular motion. Since the orbit of this movement is constantly changing and influencing the partner star, it creates an elliptical movement. Thus, all celestial bodies, including the earth, perform in circular and elliptical motion and their finite mode of existence in three-dimensional spherical motion.

2.1.1.1.3. Universal Primary Energy

Energy is the basis of everything. Without knowing anything about its origin, we can only end up, like Roger Penrose, in cyclical universes or three- dimensional universes with a star that has collapsed in the Pre-Big-bang. This is why, until his death, Fred Hoyle didn't want to hear about Big-bang and stubbornly stuck to his model of stationary universes.

To exist as an eternal and active entity, God needs energy. It is this energy contained in God that is the source of all energy. It is called the Original Energy, called the "Universal Prime Energy" by the one who discovered this notion, namely, the Rev. Dr. Sun Myung Moon. (Cf : Présentation du Principe divin : ed. Culture et Paix, Paris, 2013, Pp 29-32).

The study of the dual characteristics of God and Universal Prime Energy is a doctrinal teaching of a church but has nothing to do with doctrine, it is purely scientific. Rev. Dr. Sun Myung Moon conceived it using these notions as an engineer. It is these notions that make the greatness of Rev. Dr. Sun Myung Moon's Principle of Creation in relation to all current scientific theories and the greatness of the Divine Principle in relation to all religious teachings.

Even though this notion is part of the general teaching within the church of Rev. Dr. Sun Myung Moon to explain the origin of all things, as an economist and amateur cosmologist, I had said, I never understood it well and that is why I did not want to dwell too much on it.

The transition through beings, from virtual particles to material, is described biblically by the image verse "...he breathed into his nostrils a breath of life and man became a living soul" (Gen 2-7). Given the intellectual level of his time, Moses used man as a case study to enunciate the great laws of the Universe, which we have scientifically reconstructed in Volume I.

Thus, according to the Principle of Creation, the UPE (Universal Primary Energy) is the origin of all the energies and forces that allow created beings to exist. Thanks to the action of the Universal Primary Energy, the subject and object elements of each entity form a common base and interact according to the scheme we have presented, which solves all the mysteries of creation.

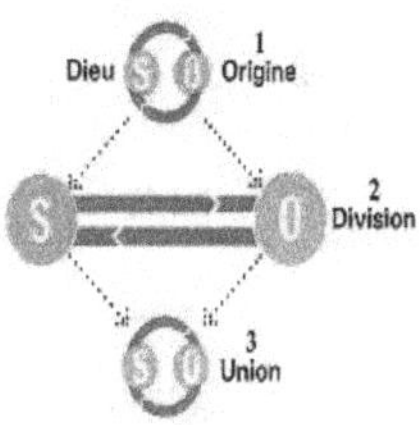

Source : The Divine Principle, Level 4, p.24

In the created Universe, the dual characteristics that make up each being receive the power of the Universal Prime Energy to establish a common basis. They then begin an action of giving and receiving.

This relationship in turn generates all the forces that the entity needs for existence, multiplication and action. The interaction that thus generates these forces is called, in the Principle of Creation, "action of giving and receiving".

Based on the forces generated by this action of giving and receiving, according to the Principle of Creation, the dual characteristics establish a foundation for their eternal exchange. This is the foundation on which God can exist eternally and generate all the forces necessary to create and maintain the Universe. (Cf : Presentation of the Divine Principle : ed. Culture et Paix, Paris, 2013, Pp 29-32).

As we have said, the literal understanding of all this is very complicated and requires a lot of schematics. Let us not forget that it is in this point of dual characteristics of God that all the mysteries are hidden; not only of the Universe, but of beings as well. It is enough to understand these notions and to apply them in the entity we are studying. What interested me the most was the fact of having understood the origin of the unbridled waltz of elements in space, starting with the spins of particles and what science calls graviton and gravitation.

BIBLE

Gen1 :2 "...the spirit of God moved over the waters" (Louis Segond). Gen 1:2 "...a wind of God moved over the waters" (B. of Jerusalem). SCIENCE

Documentary : *"New theory of the unified field : explanation of the Universe Dynamics (first part)".*

What is meant by the revolve of the Universe? This is a great debate for the supporters of the finite and infinite Universe. But by saying that the Universe rotates, we mean that the elements contained within it, as a result of the Universal Prime Energy that is endowed to them, the opposite elements of the same nature, attract each other, creating rotary and circular movements. For someone who is outside our Universe, and whose particles are visible to him, he will see himself in front of a bubble swirling like a swarm of bees grouped around a young unfertilised queen.

The globular cluster Omega Centauri seen from the ESO VST1 © ESO/Inaf-
VST/OmegaCAM, CC by-nc 4.0

futura-sciences.com/buildsv6/images/mediumoriginal/5/1

I have drawn the inference that everything in the Universe goes in circles and tends to return to unity, from the studies made on the writings of Rev. Dr. Sun Myung Moon in the Principle of Creation. This is the causal effect of the dual characteristics of God.

In principle this rotary motion originates from the spins of the particles from which every natural body is made. The particles reflecting the dual characteristics of God move on the basis of the Universal Prime Energy contained within them.

Thus, the Universal Prime Energy is the fundamental energy of God endowed in the particles. In God, all attributes, called dual characteristics, are united. Projected by God and separated, they tend to return to unity. It can be positive (masculinity) or negative (femininity), which creates the force of attraction between two bodies of the same nature, but opposed (Coulomb's Law or duality, which is none other than a natural law whose origin is in God).

This force manifesting itself within God, assuring his eternity, and within the particles for their bonding, is named Universal First Force by the Rev. Dr. Sun Myung Moon who discovered it. It is non-quantifiable and is different from potential or kinetic energy which requires a certain external force before it manifests itself.

This point, which is explained in detail in the Principle of Creation, we have also developed it in Volume I of the books in the series "The Verses of the Scholars".

According to the Principle of Creation, God is the primary cause and subject whose dual characteristics of inner form, or invisibility, and outer form are in harmony with the dual characteristics of original positivity and original negativity. In a word, God is the harmony of all characteristics. But, in religion, God is love. In the Bible in Romans 1 :20, we are told that these characteristics are clearly seen through his creation, for the latter is the emanation of his projection. This is confirmed by this quote from the physicist Richard Feynman :

"*All things, including ourselves, are made up of positive and negative parts that are closely interrelated in a powerful interaction, and all finely balanced* ". (Cf. : The physicist Richard Feynman in "Lessons in Physics").

Hence the table below

God (E.P.U.)	Particle (science)
harmony	spin
love, spirit of God	gravitation

But the driving force behind all this, according to the Principle of Creation to which we align ourselves, is the Universal Prime Energy (U.P.E.) which emanates from God and is the source of all energies in creation. Through this picture we understand why, instead of gravitation, Moses used the phrase "the spirit of God (a wind of God, according to some versions)" and Empedocles used the word "Love" as we have seen in Volume I of the series of books "The Verses of the Scholars".

Thus, although each individual being seems to exist independently of the others, the whole creation, according to the Principle of Creation, has its origin in the ideal of God who is the Being with dual characteristics in harmony. Each created being therefore does not exist independently but is created to exist through reciprocal relationships.

At the microscopic level, particles tend to return to unity, whose bosons sometimes serve as glues. This dynamism extended to the macroscopic level causes bodies to do the same, as scientists have found.

This has been demonstrated by the Dutch astronomer Jan Oort as we can

read :

"Between 1920 and 1930, Dutch astronomer Jan Oort demonstrated that the whole galaxy rotates in space by executing an enormous and complex rotation".

(Cf. : Damian Grint, Les mystères de l'Univers, ed. Christophe Colomb, P.119, Paris, 1985).

This is due to the fact that when one is created in space by mobile elements, one can only be mobile.

Hence this quotation :

"*The spin of a particle would be related to the quantum vacuum in the following way : it would not represent a rotation of the particle but a rotation of the vacuum (made up of particles and virtual antiparticles) around the particle. These virtual quantons repelled by the particle would have such a rotational movement either in one direction or the other, which would give the up and down spins.* " (Cf. : http://www.matierevolution.fr/spip.php ?08- Light and material, laws from the vacuum, Tuesday 14 July 2009, by Robert Paris)

This quote also confirms our theory that it is not only things that go round in the Universe. The Universe itself revolves around itself and transmits this dynamic to the elements within it.

As a result, all space is subject to gravitation despite the surge of dark energy. It is the gravitation that makes the Universe a creative factory and a laboratory. This is confirmed by the series of scientific documentaries "New Unified Field Theory : Explanation of the Dynamic Universe".

What revolves is this empty space we see, this empty space that contains the galaxies. The scientific phrase is that space expands by talking about expansion. Since we don't know the limits and we are contained within, it is difficult to notice this.

An airplane leaving Paris for New York is brought by two speeds. The rotational speed of the earth (1,100 km/h at the latitude of France), which is constant, and its own speed. We too, when walking or sleeping, are subject to the speed of the earth. Without this constancy for objects undergoing the Earth's gravitation, the plane would only have to rise straight up to Heaven, until it reaches a state of weightlessness and wait, with the rotation of the Earth, for New York to be on its axis and descend. So, it is this same constancy of the rotation of the Universe in relation to the things that are contained within it that makes us not notice this.

Let's take an example within everyone's reach ; close all the doors of a vehicle and release the flying insects inside. These flying insects will revolve from one corner to the other without even noticing that they are also being driven at the vehicle's speed of 100 km/hour. As they fly from one flap to the other, they are unaware that the vehicle has already taken them through a territory.

For another little experiment in moving the Universe, take a balloon or a covered bowl made of transparent glass. Let's put things of different masses inside that are likely to float in the air or in a liquid when we spin it. Let's fill it up and operate the balloon by spinning it at a speed that the human eye can only see the things inside and not the way the balloon spins. Through their different masses we notice that things will each acquire their own orbit and rotational speed. Let's blow up a mass that is inside, a black hole phenomenon is created because of the empty space left by the thing that has exploded. We have partially recreated the movement of objects within the Universe.

https://trustmyscience.com/un-trou-noir-supermassif-denude-parcourt-l'Univers/
by Stéphanie Schmidt on 4 November 2016.

We thus come to the same conclusion as Albert Einstein : celestial bodies exert pressure in space where they fill it and deform it. They fill it because they are in an orbit that allows them to perform their movements. They deform it because they cause the small bodies to swirl around them. This creates stability in space.

2.1.1.1.4. The growth period for created beings

I have already dealt with this point in the discordance between religion and science with regard to the notion of time.

Indeed, in the Principle of Creation, Rev. Dr. S.M. Moon was the first religious to make a parallelism between the day of creation in Genesis and verse 2 P 3 : 8 which states : "...that before the Lord a day is as a thousand years, and a thousand years as a day". By this we understand that the days of creation in Genesis were not actually 24-hour days. They were rather periods.

Rev. Dr. S.M. Moon was also the first to be able to explain the biblical language of "... one evening and one morning", which are periods of growth and evolution. The first chapter of Genesis tells us that the Universe did not exist instantaneously ; it was created in six gradual periods, to which we read in the Divine Principle :

"Morning" means the period in which the ideal of creation was first realized. Evening", the period when God stopped creating. By understanding the morning that came after the evening and the night; God considered all this as one day. God indicated how long it took for his ideal of creation to be realized. The period in which God began to create until the ideal of creation is realized is a period of growth for all created beings and is represented by the night, the period between evening and morning". (Cfr : Divine Principle, Level 4, p.32).

2.1.1.1.5. The invisible substantial world and the visible substantial world

The invisible world, or spiritual world, is a theoretical reality for believers, but non-existent for science. Rev. Dr. Sun Myung Moon is the first to have really described it in these various lectures and defined it in the Divine Principle. This is why this teaching should be a part of the branches of courses in all the theological faculties.

For science, we have said, the cosmos represents the Universe taken as a whole, that is, from outer space to the organized system. In other words, it is physical space.

According to the Principle of Creation, a branch of the Divine Principle of Rev. Dr. Sun Myung Moon, the cosmos represents the whole of the spiritual world and the physical world, or the universe. Since man is composed of two united entities, namely the spirit and the body, it constitutes the center of the cosmos. Since the body is composed of the physical elements of the cosmos and the spirit of the spiritual elements of the cosmos, man is the microcosm or a miniature cosmos.

The difference between the spiritual entity and the physical entity is that our physical body is limited by time, it cannot transcend the present moment. Our physical body is an ephemeral being, it grows old and returns to the earth. Our spirit is not limited by time ; no temporal barrier can stop it. It can freely reflect on the past and contemplate the future if it wants to ; it is eternal.

It is this faculty that, through reflection, two beings who do not know each other and neither of whom has read the other's work, manage to come up with the same new idea.

In our epigraphy, when the French astrophysicist and philosopher Aurélien Barrau said that with the Big-bang, Genesis was recreated in scientific terms, he did not know that one day the Big-bang would be proclaimed as biblical. When Nicholas Copernicus came to the conclusion of heliocentric, he did not know that an Arab named Nasir El din Tusi (1201-1274), a Persian scholar, had come to the same conclusion in an unpopular work several centuries before him. *(Cf. : Magazine Ciel&Espace n°413, 10/2014, Pp 10-14).*

When Father Georges Lemaitre was struggling with his Big-bang theory, he was unaware that the Russian astrophysicist Friedman had come to the same conclusion. When Gabriele Veneziano found that the Universe began in a vacuum, he didn't know that several centuries before, Moses had also come to the same conclusion which is recorded in the Bible.

An astrophysicist somewhere in the world also finds himself through what I develop but what he lacked was the knowledge of the Principle of Creation to publish these ideas before me. It was the same for the Belgian astrophysicist, Abbot Georges Lemaitre. As a religious man, he knew that if it was God who created the Universe, it could only begin with a ball, but he lacked the scientific arguments to publish this.

As soon as the American astronomer Edwin Powel Hubble discovered the existence of other galaxies and that they were moving away from each other, it was a godsend for the Belgian astrophysicist, Abbe Georges Lemaitre, to publish his theory of the original atom, which was baptized the Big-bang theory by the Russian-born American, George Gamow.

By synthesizing the discoveries of Tusi in the 13th century and Copernicus in the 16th century, the author of the article asked himself the question : "is Copernicus the last astronomer of the Maragheh school where he learned about heliocentric? Or should we humbly admit that the most daring of the theories can germinate independently in the meanders of a single man's mind". (Cf. : Magazine Ciel&Espace n°413, 10/2014, P.14).

I had evoked this coincidence of thoughts between two researchers in the foreword of Volume I.

By publishing a scientific idea, we are not closing it, but rather offering a missing link somewhere to enrich a model. As the French astrophysicist Aurélien Barrau said in Big-bang and Beyond "The birth of the Universe only happened once". Therefore, in the end, only one model should remain. And, the model of the strings, considered until then valid to explain the beginning of the Universe has, in fact, without astrophysicists knowing it beforehand, joined the Bible as restored in the Universal Cosmological Model. Therefore, as we have said, the Universal Cosmological Model has just swallowed up all the models and left the only model that explains the beginning of the Universe.

A human being is born in only two ways, either by

Caesarean section or by the lower way called the natural way. So why set several models for the same universe. All the other models become, to this day, the manifestations to explain how things happened in the universe after its formation from the quantum vacuum.

2.1.2. Quantum mechanics

We have said that it is through quantum mechanics that we study the behavior of particles that already have within them the properties derived from the dual characteristics of God. According to the action of origin-division-unit, this one seeks its opposite to initiate the actions of multiplications that lead to the formation of the elements.

The book "The Verses of the Scholars" updates the biblical energies that led to the Big Bang. Even science, in the quantum mechanics domain recognizes that at the beginning of the history of the cosmos all was energy. To speak about the creation of energies after the Big Bang as scholars try to speculate in the Standard Model does not stand. The problem is rather the place of quantum mechanics in the history of the Universe.

Some scientists prefer to put it in the sphere of the Unknown Period of the Universe as we can see in this table :

Scientific representation of the unknown period

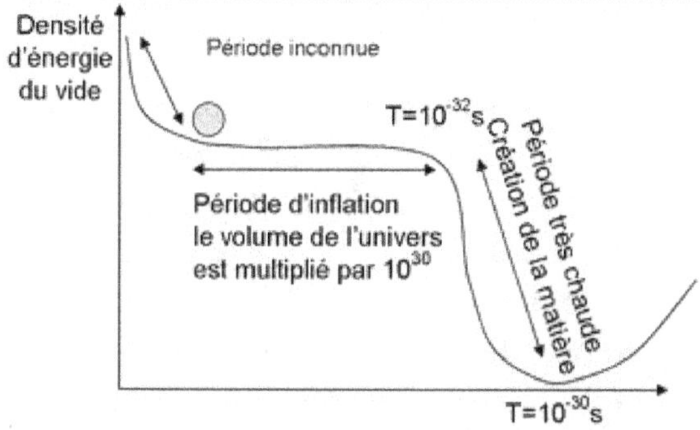

L'énergie du vide pourrait avoir évolué comme une bille roulant très doucement sur un plateau de haute altitude et en pente douce avant de chuter brusquement le long d'une corniche et de se stabiliser à sa valeur actuelle.

ces.org/articles/cosmologie-fastoche-33

Based on this table, several hypotheses have been made up concerning the uncertainty principle in order to explain the appearance of material after the Big Bang. This is the case with the following sentence :

"Even when the universe was created, while the material did not yet exist, the universe was agitated with a frantic succession of creations and disappearances of particles of all types. Moreover, it is thanks to this that ordinary material will be able to make its apparition. (Cfr:annuaire-digg des.coursgratuits.net/c7_bigbang/p715_partvirtuelles.php. By Olivier Esslinger).

For the Universal Cosmological Model, the principle of uncertainty - namely the quotation above should be deepened and positioned in the before Big-bang, that is to say, from the first universal energy up to ' the appearance of material. Everything that happened during the before Big Bang period was just a continuation of the deconfinement of what has been confined.

2.1.2.1. The notion of emptiness

Ge 1 : 2 " The earth was formless and empty ".

When we talk about the notion of emptiness, we are referring to the time when the universe (the earth) was still formless and empty according to the Bible.

One of the questions that always intrigued me in the first verse of the Bible was how the emptiness was arrived at from the spiritual, and from the vacuum to the material. In the previous volumes I had mentioned several times the speed and collision of rays. I had no idea that 10 years later I would find such an article on the Web :

"When the right frequencies of light collide, they can be transformed into material thanks to Albert Einstein's famous equation $E = mc^2$*," explains Alberto Dominguez, astrophysicist at the Complutense University of Madrid».* (Cf. : trustmyscience.com/telescope-close-trace-history-stellar-in-the-world/by Thomas Boisson,30 /11/ 2018).

Indeed, the principle of scientific uncertainty implies that even the most absolute vacuum is in fact settled down with a myriad of particles and virtual antiparticles. Emptiness, as we usually understand it meaning the absence of all material as they do not exist as such. This is the ascertainment of scientists. (Cfr : astronomes.com/le-big-bang/particule-virtuelle/ August 9, 2017 by Olivier Esslinger)

According to science, vacuum is a state that preexists material and the movement that results from it. According to general relativity, the material- energy contained in the universe is not without induced effects, because it curves space-time in the latter. Moses called it an abyss : "there was darkness on the surface of the abyss." This is why quantum physics explicitly considers without referring to the before Big Bang together with space and time as pre- existing entities. In principle, it only confirms the Bible through its scientific restitution in the Universal Cosmological Model.

Ge 1 : 2-3 "The earth was formless and empty : there was darkness on the surface of the abyss...".

In addition, Quantum Theory has shown that the vacuum was not really empty and that it contained a fabulous amount of energy, thus, the latter was then called "quantum vacuum". This void is filled with virtual particles that materialize briefly when energy is supplied, for example during high-speed particle collisions.

This "empty" state is indeed very different from a total absence. Hence, the term quantum vacuum used by science may therefore not belong to our universe according to the definition given to the latter. Hence this conclusion :

"The quantum vacuum, being filled with something could according to this definition not belong to our universe. It could for example constitute the raw material from which our universe was created 13.5 billion years ago ".(Cfr : zevengeur.wordpress.com/2014/02/11/The quantum vacuum : a new continent to explore).

This conclusion confirms what we have said, namely, in the Universal Cosmological Model that quantum mechanics with all the dances it offers us is not part of the post-Big Bang Universe. There is no place where you can insert these dances after the Big Bang characterized by inflation followed by an exponential expansion. Rather, it is a part of earlier Big Bang Universe, while according to the Bible "the earth was still formless and empty" meaning the early Universe.

When I came to this conclusion, I didn't know that I have either joined science or that the latter did the same as we can read it in the following extract :

"Contemporary physics has not answered all the questions that it searched. Indeed, while it has made it possible to study many phenomena, the laws that it has highlighted do not meet in the areas where they are supposed to act jointly or concurrently. Essentially (but not only), quantum physics does not join relativistic physics". (Cfr : matierevolution.fr, Dark matter, dark energy : the man-quant link ? 11/13/2007, by FaberSperber, Robert Paris).

Another strong argument is pointed out from Albert Einstein quotation, namely :

"For a billion particles of antimaterial, there was a billion and one particle of material. From a mutual annihilation, there is one billionth left : our present universe. "(Cfr : J. Baker : 50 keys to understanding astronomy, éd. Dunod, Paris, p.71).

(Cfr : J.Baker : 50 clés pour comprendre l'astronomie, ed. Dunod, Paris, p.71).

This is what effectively confirms that the quantum mechanics, which according to us belong to the physics of creation should be inserted in the earlier Big-bang period.

Indeed, this quantum vacuum in quantum physics corresponds to an area of the Universe that supposed to lack material or rather in which nothing can be discerned. It is the area full of virtual particles and particles of antimaterial. When particles and antiparticles collide, they annihilate each other by releasing energy until they give rise to virtual particles in the quantum vacuum. All of these actions are described by the biblical verse The power of God was moving over the water".

Today, the Universal Cosmological Model is presented as the diving suit that allows scientists to go through the arsenal of the Big Bang and to go back to the period before the Big Bang. However, since there was no star as a reference point, this period is not quantifiable either by the

Bible or by science. Moses, the scientist of scientists, must have contained in one day this period which goes from the moment when there was only emptiness and chaos to the moment when the heat of the elements that caused the Big-bang reached a great intensity, caused projecting light. Moses describes it as follows :

Ge 1 : 2-5" *The earth was formless and empty : there was darkness on the surface of the abyss, and the spirit of God moved over the waters and God said : Let there be light ! And there was light...so there was evening and there was morning, and the first day".*

For these verses we can say that there was a dark period in an empty Universe, filled with darkness on its surface, that is to say, energy. After mentioning the title, "In the beginning God created the heavens and the earth", where many phenomena still inaccessible to us had to pass through radiation by the Universal Prime Force from the Universal Prime Energy, Moses began to speak about creation at the moment when the virtual particles began to move with the energies; and this with the sentence "The earth was formless and empty..." : there was darkness over the surface of the deep, and a wind of God moved over the waters" (Gen 1:2 - Jerusalem Bible).

Given the truthfulness of this statement "of a formless and empty earth", meaning the universe at its beginning, and which is supported by some scientists, the Universal Cosmological Model now becomes the basic model in cosmology.

2.1.2.1.1. The notion of chaos

Chaos means, in the theory of the creation of the world established by the ancient Greeks, a dark and silent abyss that pre-existed all life forms...It is worth noting that although some diagrams present Chaos as a sort of confused mass, all agree that it is the first force to have invaded the Universe" (1).

(1) "Chaos. "Microsoft® Encarta® 2009 [DVD]. Microsoft Corporation, 2008.

Through this definition of the word chaos, we rather fall back into the biblical notion of the abyss, which also means dark abyss and which is defined according to the Free Bible glossary as a mysterious World of the primitive ocean, place of darkness and evil powers, of distress and perdition.

It is understood that the notion of chaos refers to what happened in the biblical abyss and whose manifestation corresponds to the "black hole" or distortion of space-time. This is why several astrophysicists who think in terms of the Big Bang speak of a star that collapsed forming a black hole from which our Universe would have come. As we have said, it is the Italian astrophysicist Gabriele Veneziano who offers an adequate scientific explanation as we can read :

"*In my scenario, we start from an almost empty, very, very diluted Universe with very weak interactions between the components. In this near-emptiness, however, there are instabilities that can become catastrophic and lead to gravitational collapses like a black hole. And the Universe as we know it would come out of this initial near vacuum through this black hole.* (Gabriele Veneziano, interview by Guy DUPLAT).

(Cf. : http://www.lalibre.be/archive/qu-y-avait-il-avant-le-big-bang ;the 15/12/2005).

Indeed, according to general relativity, energy has mass ; therefore, it gravitates. But gravitation is attractive.

(Cf. : https://www.futura-sciences.com/sciences/univers)

Thus, as it gravitates, it already digs into space and time. And, in an empty universe, this can only correspond to a collapse, as Gabriele Veneziano said.

2.1.2.1.2. Quantum vacuum and chaos

Chaos and emptiness constitute, according to all religious, philosophical and metaphysical thoughts, the state from which the pre-Big-bang Universe was born. They are found under several synonyms such as nothing, nothingness, abyss and abyss.

Ge 1 : 1-2" *In the beginning God created the heavens and the earth. The earth was formless and empty; there was darkness over the surface of the deep, and the Spirit of God moved over the waters".*

Sura al-Baquarah, 117 "*Allah is the Creator of the heavens and the earth from nothing...".*

All physicists know that fundamental physical phenomena take place in the quantum vacuum. This is what is known as "the fluctuations of the vacuum", where the appearance and disappearance of so-called "virtual" particle and anti- particle pairs take place. Thus, the vacuum is made up of dipoles (particle and antiparticle). A charged particle is constantly in electromagnetic interaction with the virtual particles that surround it. These are cycles in which the particle/antiparticle pair is transformed into energy.

In this verse; "the earth (Universe) was formless and empty", I asked myself, as I had said, the questions : where did material come from? How did the invisible become visible? I have already answered this question several times. But the conclusions based on the observations obtained from Gérard Mourou's experiments edify everyone. In these experiments, Gérard Mourou had come to the conclusion that, the vacuum is very full of particles and antiparticles (identical to particles but with opposite characteristic quantities) and all material comes from the vacuum.

In fact, this sum of material and antimaterial, which exist in a fugitive way in the vacuum and combine ceaselessly to give energy, give the overall impression that the vacuum is an absence of material. Material and vacuum are made of the same materials and material is the product of a structuring of the vacuum.

Hence this quotation from Victor Weisskopf : "*Particles only come into existence in the ordinary world through a process of creation-annihilation in the fullness that is the vacuum.*» (Cf. : http://www.matierevolution.fr/spip.php ?article38Les contradictions des quanta, 02/06/2007, by FaberSperber, Robert Paris)

In conclusion, emptiness, material, light link the Pre-Big-bang and the after Big-bang. For FaberSperber, it is the vacuum that gives birth to the material/light universe as we know it, adding that the functioning of material/light can only be understood through the permanent interaction with the 'vacuum'. It is these interactions, he says, that make it possible to understand both the laws of quantum physics and those of relativistic gravitation as well as what connects them, the quantum physics of fields.

(Cf. : http://www.matierevolution.fr/spip.php ?article43,Levide destroyer/ constructor of material, 01/07/2010, by FaberSperber, Robert Paris).

2.1.2.1.3. Energy in a vacuum

In common parlance, emptiness represents a kind of pure space, devoid of all material. In quantum physics, on the other hand, it has been found that this pure space, called the vacuum, contains energy, which quantum physicists call the "quantum vacuum". This energy is composed of virtual particles. Thus, when talking about the quantum vacuum, we refer to the pre-Big-bang as described in the Bible, namely "the earth (Universe) was formless and empty. (Ge 1 :2)

Ge 1:2 "*the earth was formless and empty*".

When we speak of vacuum energy, we are referring to the energy on the early Universe at the time when the Universe (earth) was still formless and empty, according to the Bible, in the pre-Big-bang period, and not to the vacuum energy which is now equated with dark energy; creating the expansion of the Universe.

Moses began to speak about the Genesis of the universe from the moment he said that "the earth was formless and empty. According to the quantum theory of fields and gravitation, the vacuum should have enormous energy, a theory supported by mathematical formulas from Albert Einstein's General Relativity.

Indeed, if the quantum vacuum does indeed contain a phenomenal amount of energy, nothing prevents part of it from being transformed into material, thanks to Einstein's equivalence relation (E=mc2). Thus, the origin of the Universe would then have an immanent cause, because the vacuum would have existed before the appearance of the Universe. (Cf. : "Discourse on the origin of the universe", Etienne-Klein-2010).

Wang Fuzhi (1619-1692) wrote : *"In reality, the principle resides in energy and energy is nothing but principle; energy resides in the vacuum and the vacuum is nothing but energy."*

(Cf. : JP LUMINET in Futurascience, l'infini de la matière : le vide, published on 25/01/2018).

As the earth (the universe) was initially formless and empty, material as we see it today was formed from the energy contained in this vacuum and whose properties come to us from the dual characteristics of God who has the property of bringing everything back to unity.

All this is well explained by this scientific quotation :

"*The energy of the vacuum is not null, it is constant at every point in space-time, which makes it possible to respect the law of conservation of energy throughout the universe. The energy of the vacuum is constant, but fluctuates at this value, and it is this instability of the vacuum that allows the creation of infinitely small material*". (Cf. : Claude Sarrailh, Le secret de la vie, ed. Librinova, Paris, 2018, P.42).

According to the uncertainty principle, the collision of virtual particles creates particles when energy is supplied. The collision of these pairs of particles and antiparticles can only find their existence from what is developed here.

Until the appearance of this book, The LHC has not yet reached the phenomenal speed at which the Universe spins as we can read :

"You cannot reproduce in the laboratory the acceleration of free collisions or any other particle acceleration process. We do not know how to reproduce the great distances and low densities of space, and phenomena cannot be properly miniaturized to laboratory dimensions". (Dr. David P. Stern Mail to Dr. Stern : education ["at" symbol] phy6.org).

Even particles so far subjected to the speed of light for their collisions up today will not be able to generate the original material referred to as "water" by Moses in the Bible.

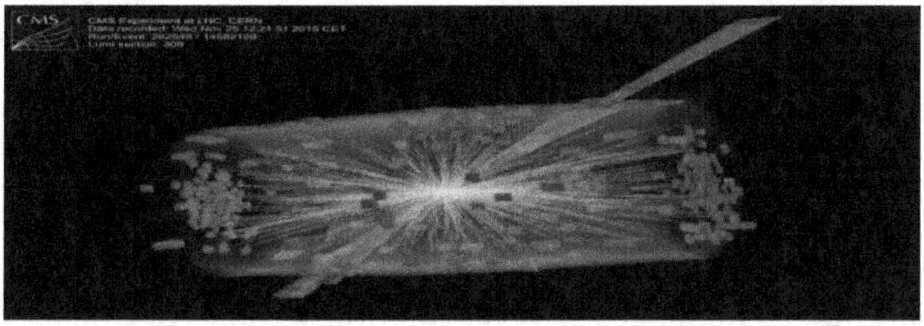

Lead ion collisions in the CMS detector (Image : CMS / CERN); to determine the state of material of the Universe at its origin

2.1.2.2. Radiation

From the spiritual to the physical and from the virtual to the real, things must have happened precisely at the level of radiation. This is what motivates us to be able to talk about it.

" *Without quanta, there would be neither light nor material and, if we may paraphrase an evangelical text, we can say that nothing that has been done has been done without them.* « (Cf. : Broglie (Louis, duc de), la Physique nouvelle et les Quanta, Paris, Flammarion, 1974).

As I have said, the passage from divine radiation to an empty Universe, described as a formless and empty earth by the Bible, has until now been a spiritual material because our apparatus does not allow us to carry out such experiments. But when we reach the level where it was described as a formless and empty earth or an empty Universe at its beginning, we enter into physics. And physics only accepts what is rational. Hence, for Louis de Broglie, everything started from quanta, the image of a formless and empty earth in relation to the heavens (God created the heavens and the universe).

Moreover, in the Bible, in the book of Romans 1:20, it is written that it is by studying creation that we can understand what is invisible in God. In the Principle of Creation, we have based ourselves much more on the examples of macroscopic creation because it is comprehensible to everyone. But, quantum physics not understandable to everyone, according to Richard Feynman, comes more and more to the origin and brings out the divine characteristics hidden, but understood by those who have mastered the physics of creation (Principle of creation plus quantum mechanics).

Thus, all notions of quantum physics are associated with radiation. One cannot study the quantum without referring to radiation. These two notions converge for some and diverge for others in what is called wave and corpuscle. Only a return to the source allows us to clarify it.

To come back to our conclusion, because there is a very abundant literature on these notions, physicists in their experiments have found that light with sufficient energy can produce material and that shocks of material produce light (a wind of God moving on the waters, there was light !).

Other experiments, called exchanges of motion, have been carried out by astrophysicists between material and radiation or radiation and radiation, between material/vacuum, radiation/vacuum or vacuum with itself. They realized that the vacuum can materialize, that material can produce radiation. Without going into detail, this really sums up everything that happened on the first biblical day as given in the Universal Cosmological Model.

2.1.2.2.1. Radiation from particles and antiparticles

According to the findings of physicists, radiation can, if it is sufficiently concentrated in energy, produce particles of material. They have discovered another kind of material, partially opposite to ordinary material, called antimaterial. It differs from material in that the charges are opposite but the mass is the same. But what changes is that when they meet (both massive and at slow speed), they produce radiation.

Previously it was believed that material and antimaterial completely annihilate each other, but finally it was found that electron and positron give the photon. Quark and antiquark give the meson Pi or Pion. And the photon of light decomposes inversely into material and antimaterial. Material is therefore everywhere, including light. This explains why material absorbs light. The thermonuclear explosion of stars shows how material can be transformed into light.

(Cf.:http://www.matierevolution.fr/spip.php ?article38Les contradictions des quanta, 02/06/2007, by FaberSperber, Robert Paris).

This paragraph confirms our theory, according to which, starting from divine radiation, it was finally able, with a decreasing deflation speed, to be transformed into physical energy. This is the consequence of observations where physicists have found that radiation can, if sufficiently concentrated in energy, produce particles of material.

In the biblical scheme of creation, restored in the Universal Cosmological Model, we are told of the vacuum (a formless and empty earth) formed on the basis of radiation from the Universal Prime Energy, and of the material represented by the waters, which, in the end, ended up again with radiation (God says that light be and there was light!). And this has led to the conclusion that material, light (radiation) and emptiness are built on common foundations. This confirms the pattern of biblical creation as reflected in the Universal

Cosmological Model. Therefore, as we have said, the Theory of the Whole can only be reached, as we have said, by taking into consideration this latter Model.

2.1.2.3. Material and Antimaterial

Just as material is made up of particles, antimaterial is made up of antiparticles. Indeed, it has been observed by scientists, as we have said, that with every elementary particle there is associated an antiparticle, carrying opposite charges, but possessing the same masses. They have similar properties including the same spin and duration.

Scientists have discovered that when material and antimaterial come into contact, an explosion occurs, generating deflagration and annihilation. Eventually, dust is formed but cannot be seen with the naked eye. Scientists speak rather than this can be followed by the emission of radiation of new and different particles. In the end, the Universe had to be null, that is to say, to have no elements, as we can read in the following quotation :

"*In the primordial Universe, particles and antiparticles constantly appear in pairs, only to be annihilated and become energy again.* « (Cf. : Blandine Plucher, la physique quantique pour les nuls, ed. First, Paris 2018, p.112).

The question : "how material prevailed over antimaterial" has always remained an enigma despite the explanations proposed by science. But with this approach, which goes hand in hand with the conclusions of the

French chemist Antoine de Lavoisier, who formulated the law of the conservation of material or mass : "nothing is lost, nothing is created, everything is transformed", scientists can get close to the results.

2.1.2.3.1. Material as concentrated energy

For observations of what is said here, not only are physicists observing the materialization of light energy in the vacuum, but it is beginning to be realized experimentally, as demonstrated by the physicist Gérard Mourou of the CNRS in Palaiseau. He has recently designed a high-energy laser beam (coherent light).

Gérard Mourou's experiment has shown that light transforms into material and he concludes that all material comes from the vacuum, that "the vacuum is the mother of all material. (...) It contains a gigantic quantity of particles per cubic centimeter... and just as many antiparticles. "His experiment was commented on in the press of the newspaper Le Monde on 19 October 2005.

(Cf. :http://www.matierevolution.fr/spip.php ?article38, Les contradictions des quanta, 02/06/2007, by FaberSperber, Robert Paris)

All this shows that material, light and emptiness intertwine and that material is a product of... emptiness. Hence the question of how, from a formless and empty earth, which was the Universe at its beginning, we end up today with a Universe filled with material ? Even if religion tells us that God is the creator, this has been in the shadows. In reality, it was these dual characteristics, contained in the virtual energies, led the latter to be transformed into material.

Today science, through quantum mechanics, is showing us that material started from the vacuum, confirming that the Universe had a beginning and that in that beginning there was nothing (the earth was formless and empty) as Moses says in the Bible. Note that these virtual energies were already contained in the vacuum or in the formless and empty earth (Gen 1:2).

2.1.2.4. Celestial mathematics

2.1.2.4.1. The philosophy of celestial mathematics

We spoke about celestial mathematics in Volume II. Here, we only want to come back to these great notions, especially since many will read this book before the previous ones.

Celestial mathematics, we said, comes from the culmination of our understanding of the Principles of Creation, a branch of the book of the Divine Principle by Rev. Dr. Sun Myung Moon. It is part of the Physics of Creation, a branch of physics created by me and which encompasses all the branches and all the ideas and theories about emptiness, as well as the physical processes taking place at the time when the earth was still formless and empty according to the Bible, that is, the beginning of the universe in scientific terms.

These equations, as we have said, are solved differently from mathematical equations because here it is the entities and it is what is invisible in them that is brought together.

For example :

– In normal mathematical equations : 1-1 = 0 or zero

– In celestial mathematics, +1-1 is inadequate, because entities do not contain the same number of positive and negative elements within them. On the other hand, their subtraction creates a dysfunction and not annihilation.

Our example refers to a Christian home in the following way, according to the following verses :

Gen 2 :24 " This is why a man leaves his father and mother and bonds with his wife, and they shall become one flesh".

– Thus, in a Christian marital union, 1 + 1 must result in 1.

Since this is an equation that requires a share of responsibility on the part of the spouses, it reads : delta 1 + delta 1 = 1.

In this equation, we aim at what is invisible to the naked eye, such as characters. The spouses must strive to unite their characters so that this does not become the cause of separation. On the contrary, they must harmonize all the time to keep 1 as an answer. Hence the formula a + b = ab instead of a+b ; which means 1+1=1.

– In Christianity, we speak of God the Father, God the Son and God the Holy Spirit as a whole. Thus, their addition gives1 + 1 + 1 = 1.

The answer 1 represents what is inside them, namely Christ, hence the verses "whoever has seen me has seen the Father" (Jn 14 :9), "I and the Father are one" (Jn 9:10), "I will send you the Holy Spirit, who will speak to you of what is mine and tell you of things to come" (Jn 14:26).

Even though their total adds up to 1, each one is an individual. Hence the biblical verses : "Jesus is seated at the right hand of the Father" (Act 7:56), "but that the world may know that I love the Father, and that I am doing as the Father has commanded me, stand up and let us depart from here" (Jn 14:31), " But the Counselor, the Holy Spirit, whom the Father will send in my name, will teach you all things and remind you of everything I have told you " (Jn 14:26). This is to say that people should not expect a Counselor with the face of Jesus or the Holy Spirit who inspires pastors.

2.1.2.4.2. The outward form of God

In the study of the dual characteristics of God contained in the Creation Principle, a branch of the Divine Principle of Rev. Dr. Sun Myung Moon, we have said that all the characteristics masculinity-femininity, positivity-negativity, invisibility (inner character) and mass (outer form) are united and are in harmony in God. Being projected into creation, they tend, according to the Principle of Creation, to return to unity, while creating multiplication and development. Hence the following verse : *" This is why a man leaves his father and mother and bonds with his wife, and they shall become one flesh*". (Gen 2 :24).

In the man, the aspects of positivity are found in an active way, while in the woman it is the aspects of negativity that are dominant. But these aspects are in harmony in God. This gives us the following heavenly equation :

Divinity = In [(Ci .M) + (Po.Ne) + (Ms. Fe)].

In : Invisible

Ci : Interior feature

M : mass

Po : positivity

Ne : negativity

Ms : Masculinity

Fe : femininity

Therefore, all external forms and their attributes are contained in God, but in a harmonious way. Harmony can only be created by a perfect turn. A perfect turn is one in which the speed between two points is achieved almost simultaneously. This speed is the **In**. This is matched by the biblical phrase "*God is everywhere and at the same time*" (Mt 6:18).

Now, to be everywhere and at the same instant in the physical world, we have said, one must rotate at the speed of the dimension of the Universe per millisecond. This is the variant value of **In**, since we do not know the limits of the dimension of the Universe.

What is the equivalence between the speed of **In** and the speed that is matched in Albert Einstein's equation of $E=mc^2$?

In the Universal Cosmological Model, we have adopted this formula as the transforming force that begins in **In**, passing through the virtual energies in an empty Universe (the earth was formless and empty) to the completion of the mass. It is this last part which concerns the E=mc2 which is considered as the absolute limit in science.

We know that the speed of light in the vacuum is about 300,000 km/s, which is 7 times around the earth in a single second. This is why the appearance of light always seems spontaneous to us. According to special relativity, as we have said, this speed is a limit that cannot be exceeded by any object with mass.

So, if we planted a stake without mass in the center of the earth at the equator, with a needle of 6,380 km (which is half the equatorial diameter of the earth), and we operated it at the speed of light, after one second it will pass 7 times over each nation. Now, in the formula "Divinity=In [(Ci.M) + (Po.Ne) + (Ms.Fe)]", it is the variant value of **In** that corresponds to the simultaneous speed to reach the furthest point in the Universe from the earth.

The furthest galaxy from earth discovered up to the time I was writing this book is called GN-z11. According to the calculation made by astrophysicists when it was discovered in March 2016, its light travelled 13.4 billion years before reaching us. Its actual distance is estimated at 32 billion light years (al) from the Earth. (Cfr : Wikipedia : GN-z11).

However, 1 al = 9.461 billion kilometer. Thus, this galaxy is 32 billion x 9,461 billion kilometer away from the Earth, which results in 302,752 x 10^{18} km.

So, for the moment, i.e. while we are writing this book, the variant value of "**In**", which also represents the speed at which God as an entity rotates to control the entire observable Universe, is 302,752 x 10^{18} km/s. This speed means that God is permanently omnipresent.

– If the speed of 300,000 km/s represents 7 times the circumference of the earth and if we take it as a reference, the speed of 302,752x10^{18} km/s will represent 302,752 x 10^{18} km/s x 7/300,000 km/s, which will give approximately 7,064x10^{15} times the circumference of the earth per second.

This is how God is more than present in us and nothing escapes Him, as Jesus Christ confirmed in these verses :

Mt 10 :29-30 "Are not two sparrows sold for a penny? However, one does not fall to Earth without the will of your Father. And even your hair is all counted".

Between 300,000 km/s of the absolute speed of light to arrive at the result of e=mc2 and 302,752 x10^{18}km/s to arrive at Dv = In [(Ci.M) + (Po.Ne) + (Ms.Fe)], we can see the travelled road to move from the spiritual universe to an empty Universe. At the beginning, I myself didn't know that there was a great chasm between the spiritual and the quantum vacuum.

Moreover, the variant value of In is evolving as man's vision advances to discover the boundaries of the Universe. We have said that everything in the Universe was generated by a movement of 302. 752 x 10^{18} km/s according to the verse "A wind of God moved over the waters" (Ge 1:2). Scientifically speaking, this movement was later characterized by the collision of particles ; annihilation and fusion released more and more energy, creating particles and anti-particles. The collision cycles continued with the creation of material and antimaterial. And in the end, material eventually prevailed according to God's ideal.

At present science is still trying to understand how particles prevailed over antiparticles and how material prevailed over antimaterial. The ideal of God or the will of God does not exclude scientific explanations, and it is up to man to discover it. As for us, we have already brought our literal version in the point concerning the dual characteristics of God to Volume I.

The collision is always perpetual, caused by the inherent properties of material according to the dual characteristics of God matched in the formula :

Divinity = In [(Ci.M) + (Po.Ne) + (Ms.Fe)], led to the energetic formation that created the first deformation of the fabric of space-time, described as the abyss in the biblical scheme of creation in Ge 1:2 (there was darkness on the surface of the abyss).

According to general relativity, space is represented not as a gas or a vacuum but rather as a stretched rubber cloth. Thus, objects endowed with mass are like balls placed on this rubber while deforming it.

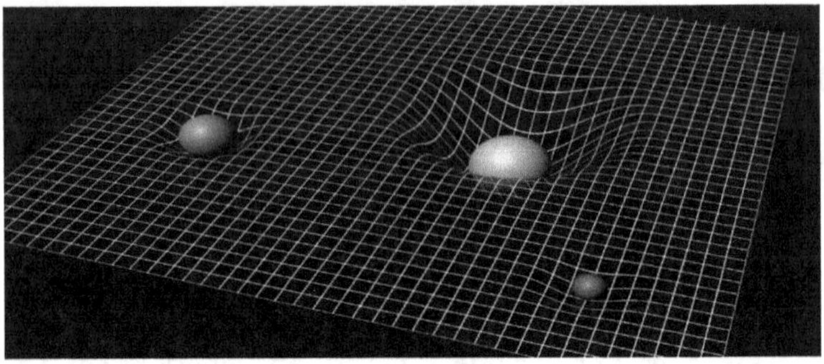

https://trustmyscience.com/reflexion-n1-lespace-temps
– **God's inner strength** (volume II P.35-37)

God's inner strength is the source of his eternity :

God = First Universal Energy

Seen from this angle, we can conclude, in the Universal Cosmological Model, that even the notion of Higgs Field falls into disuse because mass is not acquired in a field, but it goes from invisible energy to particles during their evolution. This is supported by this sentence from the Principle of Creation :

"According to modern science, all particles that make up atoms are made of energy. In order for energy to form particles, it must have an inner nature that leads it to take on specific forms too". (Cf : Presentation of the divine principle : ed. Culture et Paix, Paris, 2013, Pp 21-26).

By form, we mean mass. According to current science, particles acquire this mass by crossing the Higgs field. The slower they are; the more mass they acquire; this is referred to in the quotation above as "specific forms".

For the Universal Cosmological Model, any identifiable particle already has mass, that it can lose as a result of chemical or physical reactions. So we can say that the Higgs field, where particles acquire mass, will always remain invisible or even non-existent. Some say that it is the universe itself. However, we do not deny the existence of the particles that scientists have called Higgs bosons, which sometimes serve as links called interaction.

In fact, according to the quantum field theory (TQC), material consists of particles called fermions, while forces are transmitted by the interaction or exchange of other particles called bosons.

Our assertion that mass is intrinsic to particles actually solves this puzzle. This can be confirmed by a recent scientific article which reads as follows :

"The quarks that make up hadrons are called 'valence quarks'." Almost all of the hadron mass does not come from the valence quarks that constitute them. In fact, physicists first became aware of this with the proton when they calculated that the sum of the masses of the valence quarks - acquired through their coupling with the Higgs field - was much smaller than the measured mass of the proton. Most of the mass actually comes from the virtual gluons, due to the mass-energy equivalence". (Cf. : trustmyscience.com/hadron/ by Thomas Boisson, August 3, 2018).

Albert Einstein's formula of E=mc2 (energy = mc squared) allows us to have energy values. This formula already explains the behavior of the elements in a formless and empty Universe. But in order to get to this point, we have the original basic formula :

Divinity = In [(Ci .M) + (Po.Ne) + (Ms.Fe)],

This formula explains the addition of God's dual characteristics. By inverting each dual characteristic, we get the answer that each pair is first characterized as energy based on the original Universal First Force which is equal to the speed of radiation, the latter being equal to the speed of thought.

The slightest decrease in this speed allows the

transformation from invisible to visible energy (virtual energy), or what esoteric religions call charge. Thus, in the assertion : God = Universal Primary Energy (UPE), this force was transmitted within the entities by the phrase "God breathed" which man uses figuratively. Thus, within the entities, it manifests itself in the following way :

Di = UPE

UPE = In(FPU)

In : Invisibility

UPF : Universal Prime Force

In each entity or being, we find the energy of God called the Universal prime energy. The Projection of these entities leads them to unite according to their origin and thus generates between them an action of giving and receiving. This generates the Universal prime force. The Universal prime force, generated by the internal action of giving and receiving allows the entity to exist and be active, while creating diversification.

All this is only a part of the fundamental formulas, it is up to mathematicians and physicists to develop and enrich it after having understood the Principle of Creation, a branch of the chapter of the Divine Principle's book Rev. Dr. Sun Myung Moon.

2.1.2.5. The quantum world

This point is developed in Volume II. But in this volume we want to shed some light on it.

The quantum world or the quantum Universe is the Universe of the infinitely small, not accessible to common mortals, but which confirms to scientists the attributes of God through the behavioral study of particles. The so-called "quantum" theories describe the behavior of atoms and particles.

Scientists have been able to observe that the quantum world (limited to the infinitely small, but which can have repercussions on a larger scale) behaves very differently from the macroscopic environment to which we are accustomed. It is a milli-microscopic world. It is the subject of quantum mechanics, a branch of quantum physics. For the Universal Cosmological Model, as we have said, it is part, together with the points taken from the Principle of Creation, of the physics of creation. (Volume II P.39)

2.1.2.5.1. The dual characteristics of God and quantum mechanics (Volume II P.40-42).

Quantum mechanics is the branch of physics that studies and describes the fundamental phenomena at work in physical systems, particularly at the atomic and subatomic scale, commonly known as the microscopic scale. Thus, it is the fundamental theory of the particles of material constituting the objects of the Universe and of the fields of force animating these objects. (Cfr : futura-sciences.com/sciences/definitions/physical).

For scientists, quantum mechanics is full of mysteries, surprises and paradoxes that force them to review the way we conceive material, and even physics in general. This is because the source of everything, "God", has been neglected by science. Everything that is studied in quantum mechanics finds its cause in the extension of the omnipresence of God; matched by the formula :

Divinity = In [(Ci.M) + (Po.Ne) + (Ms.Fe)].

In = 302,752 x 10^{18} km/s : initial speed at which everything from God unfolded.

The decomposition of the highlighted elements, taken in isolation, is indeed the subject of discoveries in physics. And their behavior in order to establish theories constitute the study of quantum mechanics. With all that is written below, I am not adapting religion to physics, but rather the discovery by physical science of what is said in religion.

In fact, several phenomena studied and revealed by quantum mechanics do not belong to the post-Big-bang era; and I think so. Quantum mechanics is in fact the gateway to the transformation from the spiritual to the physical, that is to say, as soon as the word earth was first pronounced in the Bible through the verse "*In the beginning God created the heavens and the earth*" (Gen 1:1), *an earth that was still formless and empty*, that is to say the Universe (Gen 1:2) was born. It was at that moment that the aspects of God, made material in the Universe, began to return to unity according to the aspect of the Creator, thus creating the cosmic deflation that characterized the Pre-Big Bang Universe.

Therefore, the phenomena studied in quantum mechanics, made possible by a deflating and condensing Universe, do not fit with the one studied in general relativity characterized by an inflating and expanding Universe.

As we have said it above, the Nobel Prize winner Richard Feynman had the honesty to declare that "All those who claim to understand anything about quantum theory - including physicists - are liars".

The scientific commentary of this quote is as follows :

"*This means, in plain language, that the internal mechanisms that govern the behavior of particles are still completely unknown. Even if the statistical approach of quantum physics allows extremely accurate predictions, it is still only an external and clever way of describing things».* (Cfr : zevengeur.wordpress.com/la-chromo-dynamic-quantum).

We have stated that these internal mechanisms are

described in the Principle of the Creation of Rev. Dr. Sun Myung Moon, but not yet integrated into science because it is housed in a religious book given its status.

The books in the series "The Verses of the Scholars" are the first to bridge the gap between religious and scientific teachings and will henceforth allow one to enter knowledge and the other to leave ignorance.

The Rev. Dr. Sun Myung Moon who was the first to have brought out its physical characteristics in the Principle of Creation, a branch of the book of the Divine Principle, made it possible to understand the mysteries of the Universe, at least for the scholars who will master the explanations.

Through the teaching of Rev. Dr. Sun Myung Moon, we understand that the more humans advance in the understanding of truth, the more it allows God to open the vase. On several occasions Rev. Dr. Sun Myung Moon has gathered the scholars by distributing the book of the Divine Principle, which contains the Principle of Creation. But they had never understood his true intention.

With the publication of the books in the series "The Verses of the Scholars", all the so-called sacred books containing divine wisdom will now be carefully read even by atheist scholars as a research document, since faith only comes into play later after the discovery of the riches of The verses of the Scholars.

Returning to our subject, God being invisible, in physics, his mathematical characteristics have been matched by the thinkers of quantum mechanics without knowing it. This can be summed up in several points defined by D. Louapre for the7wondersofquantummechanics. *(Cf : sciencetonante.wordpress.com/2013/09/30/the 7 wonders of quantum mechanics).*

Everything that is studied in quantum mechanics finds its cause in the extension of the omnipresence of God; matched by the formula :

Divinity = In [(Ci.M) + (Po.Ne) + (Ms.Fe)].

In = 302,752 x 10^{18} km/s : initial speed at which everything from God unfolded.

The decomposition of the highlighted elements, taken in isolation, is indeed the subject of discoveries in physics. And their behavior in order to establish theories constitutes the study of quantum mechanics. With all that is written below, I am not adapting religion to physics, but rather the discovery by physical science of what is said in religion.

In Volume II, we have selected and explained some of the main points of quantum mechanics that testify to the omnipresence of God. We are not going to repeat the explanations in this volume so as not to empty the contents of Volume II, but we will only have to speak about them in a very succinct way. (Volume II P.40-42).

2.1.2.5.1.1. The principle of superimposition

When we look at things in our daily lives, including living beings, everyone occupies a specific position even if they are in motion. At a given moment, we find ourselves in a well-defined state. This is the so-called "classical" mechanics,

i.e. the mechanics of normal objects.

It is not the same with microscopic objects that undergo the dual characteristics of God. Unlike other normal objects, a microscopic particle can be in a mixture of several states. For example, according to observations, when an electron revolves around a proton, it is in all points of its orbit at the same time.

To designate the fact that in quantum mechanics, objects can be in several states at the same time, we speak of the principle of superposition. This reflects one of the divine attributes called "omnipresence", i.e. God has the ability to be everywhere at the same time. He can be active for the one called the living state and passive for the other called the dead state. Since God is invisible, it is rather the person separated from him who finds himself in a dead state while being alive. Hence the biblical verse "*let the dead bury their dead*" (Mt 8:22). This characteristic is recognized in quantum mechanics through the example of Schrödinger's famous cat, this "fictitious" cat which is both dead and alive. (Volume II P.42).

Thus the active presence of God in individuals makes one person live and the other dead; a principle which can be annotated in astrotheology by the symbolism of :

|Individual> = | Dead > + | Alive >

Indeed, to describe the two states simultaneously, i.e. a person can be both dead to God and alive to Satan, physicists have invented two kinds of brackets like this one " " and, to superimpose states, they add them together.

2.1.2.5.1.2. The indeterminism of measurement

So far we have been talking about the speed of about 302,752 x 10^{18} km/s of God's energy, for Him to be permanent throughout the Universe. However, this measurement result is potential, since we have referred to the most distant star discovered so far. If one-day science discovers more, it would change the situation but would not alter our formulas in any way. Besides, God is spiritual and we cannot give him physical values with precision; it is the indeterminism of measurement. (Volume II P.44)

Going back to our example of the electron which can

be present everywhere around the proton, physicists have discovered that each time they calculate its speed, it changes or sometimes repeats itself. This is noted by the sign :

Electron> = |1000 km/h > + |2000 km/h >

This means that by measuring the speed of the electron, one will find either one (1000 km/h) or the other (2000 km/h), but there is no way of knowing in advance which one.

2.1.2.5.1.3. The wave-corpuscle duality

We said that the coefficient of the divine speed that we used in the calculations was related to the probability of finding God at every place in the Universe, known theologically as the omnipresence of God. When transmitted to radiation, this radiation became heavier to generate punctual particles with the dual characteristics of God, from which they emanated, and which would later be transmitted to material. Hence the punctual objects according to David Louapre in the documentary "Quantum mechanics in 7 ideas" behave, at the microscopic level, like waves, i.e. they are everywhere at once. (volume II P.46).

This idea that waves were already material was introduced at the beginning of the 20th century by several researchers, including the French physicist Louis De Broglie. De Broglie proposed the idea of wave-corpuscle duality : particles can follow circumstances and behave either as particles or as waves. This idea put an end to the nature of light, which is made up of electromagnetic waves and photons.

2.1.2.5.1.4. The tunnel effect

One of God's attributes is that He has no barriers; He transcends time and space. Our physical barriers are nothing to Him. He even probes our thoughts and hearts (in the religious sense) as this verse attests :

Mt 6:6 *But when you pray, go into your room, shut your door, and pray to your Father who is there in the secret place; and your Father, who sees in secret, will restore it to you.*

The Universe is a divine projection, this attribute is preserved by certain rays and microscopic elements, which are the subject of study in quantum mechanics. If God is global but to each individual, he becomes a personal God known as the Father. (Volume II P.47)

Mt 23:9 *And call no one on earth your Father; for one is your Father, he who is in heaven.*

This characteristic acquired by certain microscopic physical elements, one slice that can transcend a barrier and end up on the other side, is called the tunnel effect.

2.1.2.5.1.5. The path integral

Believers sometimes have special places for meditation, but God transcends time and space. He does not follow a specific path to reach us and no barrier can stop him. As stated in the tunnel effect, he has no barriers.

This characteristic acquired by microscopic elements means that in quantum mechanics, when a quantum particle goes from one point to another, it passes through all the possible paths that connect these two points. This is demonstrated by several Feynman diagrams that astrophysicists are more familiar with. This approach is scientifically known as the path integral. (Volume II P.48).

2.1.2.5.1.6. Teleportation

Teleportation is the exchange of information in an almost instantaneous manner without the use of voice, gestures or external tools. It is not limited to thoughts only, but covers all psychic areas such as feelings, pains, apprehensions, etc.

Through quantum mechanics, scientists are beginning to explain, based on the behavior of particles, what was previously part of the psyche. This takes them away from religion and philosophy, but for us it is the opposite, all the more so since God is the creator and has endowed man with all the capacities through his creation.

Indeed, for the French physicist Claude Sarrailh, "*the elementary particles which are the vectors of action have no problem with time (the past, the future). Before interacting, they are everywhere in space-time and time for them does not pass... Let us remember that we are only made of atoms where electrons revolve around which we don't know before observing them where they are (they have no physical reality); And these electrons have the power to go into the future and to come back instantaneously in the present instant, since before observing them they were in all space-time, without constraint linked to terrestrial time because for them time has not passed*". (Cfr : Claude Sarrailh, Le secret de la vie, edited by LIBRINOVA, Paris 2018, Pp 95-100).

Moreover, scientists observed that the state of entangled atoms (a very complicated notion to explain in a book of literature) changes faster than the time it takes for laser rays to reach an object, i.e. faster than the speed of light. This is justifying for us, from the point of physical life, this characteristic of instantaneous deportation.

Thus, the quantum objects that make the conjunction between God and macroscopic objects keep these characteristics, even if it is decreasing as when we channel an electron and observe it. (volume II P.49).

2.1.2.5.1.7. Quantum entanglement

Quantum entanglement can be understood as teleportation at the microscopic level. According to quantum mechanics, two particles are entangled when they have a common history, here we can even say that when they come from the same atom. The entanglement means that, whatever the distance between them, their behavior is identical. A change in one of them leads, in a simultaneous way, to a change in the other. (Volume II P.51)

For Blandine Pluchet, the experiment observed during the demonstration on entanglement in 2000 called into question the principle of causality (which says that the cause always precedes the effect) because the latter is dependent on the time that elapses (Cf. : Blandine Pluchet, la physique quantique pour les nuls, ed. First, Paris 2018, p. 139).

As we can see, this finding confirms our theory of the speed of divine invisibility of 302,752 x 10^{18} km/s, which means that God is everywhere and at the same instant. And, as we have also said, quantum phenomena only confirm God's dual characteristics.

Moreover, the entanglement principle, a branch of quantum mechanics, confirms to physicists, without knowing it, the biblical and Koranic writings and the Divine Principle that make God the origin of everything. In fact, through this principle, scientists have discovered that we are all united and not separated according to appearances and this contradicts Charles Darwin's theory of evolution. (Cf. : documentary "Quantum physics for dummies"). Just as in the Bible, many of the phenomena discovered in the quantum world are still subject to interpretation.

Moreover, the entanglement principle, a branch of quantum mechanics, confirms to physicists, without knowing it, the biblical and Koranic writings and the Divine Principle that make God the origin of everything. In fact, through this principle, scientists have discovered that we are all united and not separated according to appearances and this contradicts Charles Darwin's theory of evolution. (Cfr : documentary "Quantum physics for dummies"). Just as in the Bible, many of the phenomena discovered in the quantum world are still subject to interpretation.

Section 3 : The large structures of the Pre-Big-bang

Until now, we have dwelt too much on what happened inside a "formless and empty Earth" by mixing the notions contained in Volumes I and II. In fact, many astrophysicists, without knowing that they were supporting the Bible, have reasoned and continue to reason about the emergence of material from the vacuum. Whether it's the uncertainty principle, or other principles I don't know, contributing to the emergence of real particles from the vacuum, these are theories to be put back in the Pre-Big-bang.

In Volume I, entitled "Cosmology according to biblical and ancient models", we have detailed and explained all the contents of the Pre-Big-bang as revealed in the Bible in Genesis 1:2-5 and which is articulated as follows :

Gen 1:2 *The earth was formless and empty : there was darkness over the face of the deep, and the Spirit of God moved over the waters. 3 And God said, let there be light, and there was light. 4 And God saw that, there was light good; and God separated the light from the darkness. 5 And God called the light day, and the darkness he called night. And there was evening and there was morning; the first day.*

This is the physical and rational answer of all those who ask themselves what was before the Big Bang. We have developed each set in detail in Volume

I. Speaking about it succinctly here will raise many questions because it is indeed the discovery of the great universe before the Big-bang not just words.

However, to respect the title of the book, I will give here the introduction of each set while leaving their development in Volume I which effectively reflects the great work I have done for the superiority of ancient Jewish wisdom, contained in the Bible and entitled "Cosmology according to Biblical and ancient models".

What I have developed in Volume I as part of my ten years of research has nothing to do with my status as a believer. Simply, it is my discovery of the pre- Big-bang era. I have looked in astrophysics for evidence of its confirmation, which I give you in my publications, and in Rev. Dr. Sun Myung Moon's Principle of Creation, "The Why". Namely that the "Why" is one of the answers one seeks from a creator. I am the only one in cosmology who has done this in a scientific way, that is, by probing other sacred writings in an objective way. Hence, by linking the "how" and the "why", this gives us astrotheology.

We said that it was in the Bible and not in science that we discovered that the pre-Big-bang was a very turbulent reality. Everything that happened was summed up by Moses through these verses which he called the period of one day :

Ge 1 :2 *The earth was formless and empty : there was darkness over the surface of the deep, and the Spirit of God moved over the waters. 3 And God said, Let there be light. And there was light. 4 And God saw that, there was light good ; and God separated the light from the darkness. 5 And God called the light day, and the darkness he called night. And there was evening and there was morning : and it was the first day.*

As promised, here is the introduction for each set.

3.1. The earth is formless and empty

According to the Universal Cosmological Model, the Universe began in a vacuum as the following verse attests : "*In the beginning God created the heavens and the earth. The earth was formless and empty*" (Gen 1:2).

Let us only come back to the first word the "earth". This is different from the planet earth. Indeed, while speaking about an empty and unformed earth on the first day of creation, it would simply mean that the earth itself, as a visible material element had no existence. It is on the third day (in the 3rd step) of creation that the planet earth is discovered to have been materially formed. We can read it as following :

Ge 1:10 God called the dry land earth, and he called the cluster of the waters seas... 13 so there was evening, and there was morning : it was the third day. So what does the sentence, "*The earth was formless and empty*" (Ge 1 :

2) symbolize in the Bible on the first day ? It is simply the universe at its origin. An empty and formless universe.

Mathematically speaking, Moses used the names of sets much more in comprehension rather than in extension, and in theology we speak about pictorial language based on forms rather than contents, because he wanted to demonstrate the existence of God the Creator to all people in general while writing these things, but not just only for intellectuals' class in particular.

So, when speaking of a formless and empty earth, Moses was only speaking of the Universe at its origin. Today, the word empty is debated in the scientific world, because it has been discovered that the vacuum physically is not necessarily empty, but it is filled with energy as attested by this article :

*"For quantum physicists, a vacuum is not anything : stripped of all material and at a temperature closer to an absolute zero, the empty space still contains energy! The proof of its existence? The "Casimir effect", capable of generating movement from this single energy «. (*Cfr : The force that comes from the void - Astrid Lambrecht in monthly n ° 376/06/2004, p.48).

In the Casimir effect experience, it was discovered that two conductive plates placed very closer to each other spontaneously undergo an attractive force which can be measured. (Volume I P.147)

We have fully developed this point in the notion of emptiness, chaos and radiation contained in this book to return to it. For a review, how does material emerge from this emptiness? Believers will say that God created it, but how, physically, did it happen? This is all that astrophysicists expect.

Not wanting to turn our research into a textbook of faith, we line up around the following quote :

"*The energy of the vacuum is not zero, it is constant at every point in space-time, which makes it possible to respect the law of conservation of energy throughout the universe. The energy of the vacuum is constant, but fluctuates at this value, and it is this instability of the vacuum that allows the creation of infinitely small material*". (Cf. : Claude Sarrailh, Le secret de la vie, ed. Librinova, Paris, 2018, P.42).

As the earth (the Universe) was initially formless and empty, material, as we see it today, was formed from the energy contained in this vacuum and whose properties come to us from the dual characteristics of God who has the property of bringing everything back to unity according to the law of ODU (Origin - Division - Union) issued by the Rev. Dr. Sun Myung Moon. (Cf. : Presentation of the Divine Principle, ed. Culture et Paix, Paris 2013).

3.2. Darkness

Ge 1:2" *The earth was formless and void : there was darkness on the surface of the deep and the Spirit of God moved over the waters. 3 And God said, let there be light. And there was light. And God saw that the light was good;*

and God separated the light from the darkness.4 And God called the "light" day, and the "darkness" he called night. So there was evening and there was morning, and that was the first day.

If all the stars were to go out today, including the Sun, we would return to a dark Universe as we were at the beginning. This still testifies to the truthfulness of the biblical model given in the Universal Cosmological Model. This point solves the problem of the Olbers paradox we spoke about in Volume II. (Volume I P.147)

In the third verse (Gen 1.3), Moses speaks of the appearance of light, but a light never appears, it needs a foundation or base. So what is the scientific source of this light? Here we see the appearance in darkness of a light that has no source; that is to say, a light that has as its source neither the Sun, nor the stars, nor any star, because at that time neither the Bible nor science spoke of the creation of these things. The Bible simply says that God says : "Let there be light". And there was light" (Gen 1 :3). And God called that light "day" and that darkness "night".

Since "light" in this context does not represent our sunrays, so what about darkness? Since "darkness" is a state of something that we cannot touch, we say it should be a kind of energy that preceded the Big Bang moment. By asking science to know what kind of energies are involved in cosmology; we discover after much research that it should be dark energy, which is still the topic subject of many researches by scientists till today. (Volume I P.155)

To those whom this name is quite new, we can simply say that the astrophysicists having observed certain phenomena that they cannot explain, including the acceleration of the universe, due to the distance between galaxy, have concluded that there is an unknown energy which contributes to this. They called it dark energy. Let us read this :

"Scientifically speaking, dark energy is the name given to a hypothetical component of our universe, characterized by having negative pressure. While the standard Big-bang theory, coupled with observations made so far predicted that the expansion of the universe would slow down under the effect of gravity, it seems to have accelerated over the past few billion years. This is how the idea of an invisible energy came into being, which has been renamed dark energy and which today has been scientifically proven since 1998 by two teams of astrophysicists led respectively by Saul Perlmutter and Adam Riess "(Cfr : 13/11/2015 -Laurent Sacco : Futura Science on web).

An ancient philosopher of the 5th century named Empédocle also spoke of this energy and nicknamed it hatred, responsible for discord or acceleration of the Universe. We developed it in Volume I.

In the scientific documentary "New Theory of the Unified Field (Explanation of Gravitation)" that I discovered in 2018, without referring to the Bible the author may never have read, demonstrates that everything began with the dark energy that filled the universe, and the movement on these photons was the source of the creation of some elements. (volume I P.156)

Moreover, according to the scientists, it is this energy that controls destiny, as this other scientific analysis asserts once again :

"Dark energy may also be the residual trace of an inflationary period in the very early history of the universe. If this is the case, then it is deeply linked to the very birth of the observable Universe, the material it contains and even the formation of galaxies and galaxy clusters. What is almost certain is that it must control the destiny of the Universe».

(Laurent SACCO, www.futura science, 04/10/2011).

These are comments that are in line with the Bible and support the Universal Cosmological Model.

Currently dark energy is recognized by science as a negative force which provides the expansion of the universe. There remains an enigma for the new science today which is unable neither to detect it nor explain its origin. But, as in the Standard Model, the Big Bang is considered to be the starting point of the origin of all things while scholars say that it was created during this rupture.

In the Bible, on the other hand, darkness is the seat of this dark energy, and it did exist well and truly long ago before the "separation of the water from the waters" or Big-bang. Under divine inspiration, Moses called it "night" even before the creation of the sun and stars. It means that the word, "night" in the Biblical scheme of creation has nothing to do with our present nights caused by a change in the weather. (Volume I P.158)

3.3. The Abyss

Ge 1:2 "*The earth was formless and empty, and darkness was over the surface of the deep, and the Spirit of God was moving over the waters.3 God said, 'Let there be light. And there was light.*

Artist's view of a rotating black hole (Wikipedia).

In the biblical glossary from the Free Bible, we said, the abyss is defined as "*a mysterious World of the primeval ocean, a place of darkness and evil powers, of distress and perdition*". It was from this definition that I came up with the idea of referring to black holes in the Universe.

Indeed, the glossary speaks of the word perdition, that is to say, once you are drawn there, there is no hope of coming out. And in the sky, the only things that respond to the above explanations that astrophysicists have discovered are black holes.

The concept abyss used by Moses in the Bible literally fits with what physicist John Wheeler scientifically has explained about black holes as we can read it in the following expert :

"The term 'black hole' was invented by the American physicist John Wheeler in 1967 to describe a concentration of mass-energy which gravitationally collapsed under its own force of gravity and which has become so compact that even photons cannot escape this gravitational force" (Cfr : Laurent SACCO, www.futurascience, Trou noir).

We have said that if the scientists did not bypass the biblical language, John Wheeler could have defined in 1967, the black holes of the abysses.

In space, the whirlwind of the black hole looks like this :

A supermassive black hole stripped of its galaxypassed through the universe like a whirlwind in complete freedom and at high speed by Stéphanie Schmidt, November 4, 2016, trustmyscience.com.
(Cfr:https://trustmyscience.com/un-trou-noir-supermassif-denude-parcourt-l-univers).

Schematically speaking, the force around the black hole looks like this :

They are usually represented as gigantic siphons capable of distorting the space-time around them (the blue lines in the illustration). (Cfr : Joe Bergeron, Encarta ® 2009. Modeling a black hole).

These images do indeed meet the definition of Moses who said, "...darkness was over the surface of the deep, and the Spirit of God was moving over the waters. God commanded : let there be light! And light appeared". (Ge 1 : 2-3).

The meeting of all the lines forms the bottom of the funnel in the above picture defined as the abyss by Moses in the Bible and existed when the Universe was empty. The work movement of the spirit of God or a wind of God is represented by the lines. At the bottom, we discover what astrophysicists call the wormhole. (Volume I P.161)

Above, it is the light which springs from the fusion of the original material (of the water) and which illuminates the space (God commanded : let there be light! And there was light appeared). Hence the biblical sentence : "*when he established the clouds above and fixed securely the fountains of the deep.*" (Prov 8:28)

3.3.1. The two conceptions of the word "abyss".

All those who have talked about the Pre-Big-bang say that there was a collapse before the Big-bang. For some, it was a collapsing star, or the Universe itself, and for others, like Gabriele Veneziano, it was only the deformation of space-time created by the particles combining. (volume I P.177)

He expresses his thoughts in this way :

"In my scenario, we start with an almost empty universe, very, very diluted with very weak interactions between the components. In this near vacuum, however, there are instabilities that can become catastrophic and lead to gravitational collapses like a black hole. And the Universe as we know it would emerge from this initial near-vacuum through this black hole. To be able to explain this phenomenon, we must use string theory and solve its very complex equations which involve a 10-dimensional world. For this we need new results in mathematics. To give birth to a Universe like ours, we must start from a black hole that would have at least the size of an atom nucleus (ie 10 exponents-15 m) ". (Gabriele Veneziano, interview by Guy DUPLAT).

http://www.lalibre.be/archive/qu-y-avait-il-avant-le-big-bang; published on 12/15/2005.

This Interview explains the beginning of the universe according to the Biblein a very scientific way, until ending with the notion of primitive atom of the Belgian scientist, the abbot George Lemaitre. Extrapolating with the Bible, we will have the following similarities :

ITEMS	**Gabriele Veneziano**	**Moïse** (BIBLE : Ge 1 :2)
01	Almost empty universe	The earth was formless and empty
02	gravitationalcollapseslikea black hole	Abyss
03	a black hole that would be at least the size of an atom nucleus	A wind of God was moving above the waters
04	the Universe we know would emerge from this initial quasi-vacuum through this black hole	Separation of water from water (Big- bang)

In the light of the above, we conclude that the biblical abyss is not another phenomenon than the deformation of the fabric of space-time, caused by the original material before its condensation under the action of the spirit of God (The First Universal-saddle Force), which moved above the waters (gravitational force). In other words, these verses explain physical action and must be scientifically recognized. The Bible speaks of an abyss followed by the movement of the spirit of God on the elements (waters) that it was able to gather and condense by creating, according to Gabriele Veneziano, "*a nucleus of an atom (ie 10 exponent - 15 m) to give birth to a Universe like ours*".

(Cfr :Gabriele Veneziano, interview by Guy DUPLAT on www.lalibre.be/qu-y-avait-il- avant-le-big-bang; published on 12/15/2005).

This hypothesis of Gabriele Veneziano also meets that of Stephen Hawking and Roger Penrose, as we can read : "*With his inseparable fellow student Roger Penrose, Hawking began, from 1965 to 1970, theoretical work on the Big-bang. They applied Einstein's general relativity and established that at the beginning of the Universe, there exists a singularity, a gravitational collapse similar to that found at the heart of black holes». (*Cf. : Ciel&Espace magazine n°423 - August 2005 ; The Strange Universe of Mr. Hawking HAWKING, p.42)

For the Universal Cosmological Model, the Universe was condensed before the Big Bang. But as it condensed, the material it condensed was called "water" by the Bible suite of its state. Indeed, for there to be a Big-bang, or separation of the waters from the waters according to the Bible, there had to be a zone of attraction where the original particles began to intertwine according to the literary term, creating shocks, generating heat and light, to arrive at the Big-bang. It is this zone that Moses called the abyss, and which we call the first black hole in the Universe, or the first deformation of the fabric of space-time. It is in this zone that energy (darkness) was on the surface ; not in a static way, but active, as it still is today.

In addition, the uniform temperature of the cosmological diffuse background reveals that the universe had indeed been condensed, which allowed all the extremes to benefit from the same temperature rise. (volume I P.165)

Moreover, the idea of science according to which the Universe emerged from the black hole (abyss) begins also to gain the ground as we can read it in the following scientific commentary appeared in Futura-science.fr :

"An ordinary black hole is created by the gravitational collapse of a massive star; it is a local event in the first hand. Big bang models, on the other hand are events on a global scale ; so it is difficult to relate them. Nevertheless, we start to consider models in which there is a front Big-bang. We can imagine and calculate solutions which put in scene of a Universe which was in generalized contraction for example. This Universe, instead of collapsing on itself in a Big Crunch it would have formed a black hole.

However, according to this theory we know that under certain conditions the "bottom" of a black hole is not necessarily blocked : it can produce a wormhole. This wormhole would be a kind of tunnel that could connect the entrance of a black hole to an exit : a white hole or a white fountain. It's quite tempting to imagine that our Big Bang could ultimately be the outcome out of a wormhole. This is one of the few hypotheses on which researchers are currently working". - (Cfr : Jonathan SARE, www.futura- Sciences, published on 05/11/2014).

So, scientifically, more and more conclusions are beginning to germinate that everything began in a vacuum that can be described as an abyss in space. Thus, as the first phenomena whose increasing order of emergence is not specified in the Bible, we have darkness (negative black energy), the abyss (black hole), the spirit of God (positive energy) and water (original material).

3.4. The spirit of God or a wind of God

Ge 1:2 *"...the Spirit of God moved over the waters" (Louis Segond Version) Ge 1:2 "...a wind of God moved over the waters" (Jerusalem Version)*

When reading the spirit of God, one thinks that we are talking about spirituality, but I was surprised to find that in the Jerusalem version of the Bible, we speak rather of a "wind of God". By speaking of wind, we know directly that we are dealing with a physical force, but the author attributes it to divine power. Anyone can attribute it to whomever they want, because it was the collision of particles as studied in quantum mechanics. Now, it is thanks to the dual characteristics of God endowed to these particles that they would collide. Therefore, Moses was right to say "a wind of God" according to the Jerusalem version of the Bible.

In the above verse (Gen 1:2) we see that God's spirit

was at work over the waters. It is the translation that made Bible readers think that there was water underneath and the spirit or wind of God above it. The wind in this context is an invisible force that was acting on the original particles in an empty Universe. The scientific terms nowadays are gravitational force or super force. The waters here represent the original particles.

We understand directly that the Spirit of God, or a wind of God, represents the First Energy that was in action and caused the first condensation of the Universe. For lack of a scientific name, we call this moving force on original material by the name that the Rev. Dr. Sun Myung Moon had to use, namely "the Universal Prime Force", generated by the Universal Prime Energy that is in God.

As a reminder, the Universal Prime Energy is the fundamental energy of God, endowed in particles. In God, all the attributes, called dual characteristics, are united. Projected by God and separated, they tend to return to unity. It can be positive (masculinity) or negative (femininity), this is what creates the force of attraction between two bodies of the same nature, but opposed (Coulomb's Law or duality which is none other than a natural law whose origin is in God).

This force manifesting itself within God, assuring his eternity, and within the particles for their bonding, is baptized by the Rev. Dr. Sun Myung Moon who discovered it, the First Universal Force.

The spirit of God or a "wind of God" in the Jerusalem version of the Bible is the energy that enabled material to be put into action in the period before the Big Bang. Thus, it was a very eventful period.

Since science has not yet undertaken the study of the pre-Big-bang period, then we will base ourselves on the explanation of the Principle of Creation by Rev. Dr. Sun

Myung Moon, which explains the integration of the cause in the entities, the source of their movements.

"God, the Creator of all things, is the absolute, eternal, self-existent reality, transcending time and space. The fundamental energy of God is also absolute, eternal, and existing by itself. It is the origin of the energies and forces that enable created beings to exist. We call this fundamental energy the Universal Prime Energy...through the action of the Universal Prime Energy, the subject and object elements of each entity form a common basis and interact. This relationship in turn generates all the forces that the entity needs for existence, multiplication and action. The interaction which thus generates these forces is called the action of giving and receiving". (Cf : Presentation of the Divine Principle : Pp 29-47, ed. Culture et Paix, Paris).

For science, by "subject" we mean "positive" and by "object" we mean "negative". This is how we find in common, according to the above presentation to which I refer, that everything created by God, from the atom to the compound bodies, has its positive or negative correspondent, in order to produce a multiplicative result, rather than inertia or annihilation, as we can still read in the same book :

"In this way we discover in all creation (Universe for scientists) that there is a reciprocal relationship between positivity and negativity. For example, atoms are formed from the reciprocal relationship between positive and negative elements.

Atoms themselves have both positive and negative characteristics. On the basis of such characteristics, two or more atoms maintain a reciprocal relationship, forming molecules. Plants contain male and female elements. All plants and animals exist and renew themselves by virtue of myriads of interdependent internal and external relationships of positive and negative elements...The fact that creation is composed of dual characteristics of positivity and negativity means that God himself, who is the primary cause of all things, is the origin of the dual characteristics of positivity and negativity». (Cf. : The Divine Principle, Level 4, p.11).

Through these writings of the Principle of Creation, a branch of the Divine Principle, we understand that in the phrase "the spirit of God moved in the waters", it is a force on material from Pre-Big-bang to produce after a long period of time light, which we call according to its correspondent in today's science of radiation (And, there was light). After this radiation which, we have said, marked the final stage of the pre-Big-bang. The Big-bang must have occurred afterwards as we can read in this scientific statement :

"*The maximum value of the gravitational force was reached at the moment of the Big-bang, as to say at the creation of the universe, where all material was concentrated in one point and time did not exist. It was at the moment of this release of energy and light that time appeared».* (Cf. : Claude Sarrailh, Le secret de la vie, ed. Librinova, 2018, P.15).

Without commenting on the author's idea, whose opinion of the pre-Big- bang I am unaware of, by stating that the "maximum value of the force of gravity had been reached at the moment of the Big-bang", we can see directly that there was indeed an eventful pre-Big-bang as confirmed by the biblical verse "a wind of God or a spirit of God was moving on the waters" (Ge 1.2).

This spirit of God, coming from the energy called the First Universal Force by Rev. Dr. Sun Myung Moon, contributes to the dynamics of the Universe as the fundamental force of creatures as an internal force. We find it biblically when God created man by saying in the verse : "God breathed into his nostrils and man became a living being" (Gen 2 :7).

Between two separate entities, this takes different names such as gravitation, magnetism, and all the laws of attraction and repulsion known to science. On the scale of the Universe, we also find it in the theory of the unification of forces.

3.5. Water

3.5.1. The use of the word "water" by Moses

Ge 1:2 *"The earth was formless and empty, and darkness was over the face of the deep, and the Spirit of God moved over the waters.*

The uniformity of the elements contained in the galaxies shows the biblical truth of having qualified the original particles in the formation of everything we see today, including man, water. Moreover, Moses' use of the name "water" is justified by the fact that original material, wherever it is projected into the Universe, migrates more easily into solids, liquids, gases and plasma. This is what justifies its being found almost everywhere in one of these four forms.

The form depends on the temperature to which it is subjected. For this reason, we can read this scientific statement :

"Water is abundant in the Cosmos. One could almost say that it is "humid". But it is rare to find it in a liquid state. (Cf. : The challenges of science ; the origin of life

– edited by RBA Coleccionables, S.A.U. Barcelona, 2018, P.22).

According to the Bible, the original material, called "waters", went through three phases :

– Pre-Big-bang : the spirit of God moved on the waters ;

– At the Big-bang : Separation of the waters from the waters;

– After Big-bang : formation of the earth and the lights from the waters below and above.

Thus, when we speak of water as a mass Pre-Big-bang, it has nothing to do with liquid water as we know it. We can even say that the water here represents the original particles in the mass or energies that caused the Big-bang until the formation of the structures in the post-Big-bang era.

In the light of all this and reading the story of Genesis, clarified by this book, we discover that these are truths that can be tested today scientifically.

3.6. The light

Ge 1.3:" *God says : Let there be light! And there was light"*.

The idea that a light illuminated the universe before the Big Bang seems absurd to contemporary scholars. For them, the first light only appeared after the atoms had reached a sufficient degree of heat. This is coming after the Big- bang according to standard theory. (volume I P.206)

In the Universal Cosmological Model, on the other hand, the minus one era of the universe, called the first day in the Bible, ended with the appearance of a light. Was it caused by which elements or particles?

We have said that all light has an emitting source. As for the biblical verse, no one has ever asked the question because it is written : "*God says : Let there be light! And there was light*" (Gen 1 :3). Until then, everyone except the scholars believed in a miraculous emergence of light dictated by the word of God.

Rather, it was the consequence of the movement of a wind of God, or the spirit of God, over the waters (Ge 1 :2). We understand directly that we were faced with the physical phenomena taking place in the Pre-Big-bang and that these waters, which were heated to the point of producing light, are not our running water.

Let's follow that :

Gen 1 : 2 The earth was formless and empty; darkness was over the surface of the deep, and the Spirit of God was moving over the waters. 3 God said, let there be light! And there was light. 4 God saw that the light was good; and God separated the light from the darkness. 5 God called the light Day, and the darkness he called Night. So there was evening, and there was morning-the first day.

This original first light could be the fossil radiation. Why could it be ? Because it's up to science to confirm it. We do not want to make statements based on faith in this book. Moses called it a "day."

The word "light" is biblical evidence confirmed by science that indeed the universe experienced a very high heat phase before the Big Bang. It also comforts us as a demonstration that the creation scheme is not a tale. So, let's see some properties that confirm the appearance of light before the Big Bang. (Volume I P.216)

Many may think that this radiation has disappeared since the Big-bang, giving way to fossil radiation. This cannot be true as God, in the writings of the Bible, immortalized it by the term "day"; that is to say, it has fluctuations that can no longer end. Therefore, we can say that this light is none other than the fossil radiation that remained trapped in the condensed Universe, before reappearing

380. 000 years later after the Big-bang.

The scientist George Gamow was the first to say, towards the end of the 1940s, that if the Universe had really warmed up until the Big-bang, the radiation emitted would have left traces. And it was in 1965 that two American scientists discovered them, and science called them fossil radiation or the cosmic microwave background.

This radiation would reveal that the temperature of space is strangely uniform, even in the most remote parts of the universe, and that if the Big Bang were to be taken as the starting point, the equation would be impossible. In other words, according to the Universal Cosmological Model, the maximum warming of this radiation was reached before the Big Bang.

This description of the appearance of a light before the Big-bang is supported by all the scientists who have reasoned about the Pre-Big-bang, especially when talking about a star that collapsed. As for Roger Penrose, the creator of the theory of cyclic universes, he explains it through the following diagram, called the Minkowski diagram.

Conformal diagram of Minkowski's space-time.

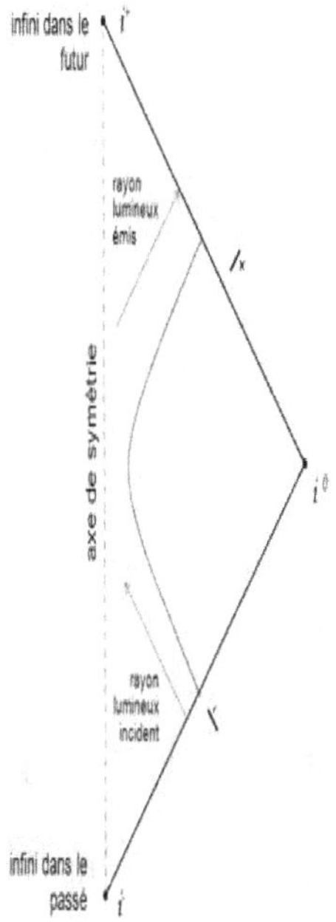

"*The conforming nature of this representation means that a light beam from the past follows a straight line inclined at 45 degrees from right to left and a light beam directed towards the future follows a straight line inclined at 45 degrees from left to right. The blue curve represents the trajectory of an object coming from infinity in the past and extending to infinity in the futur*e".(Diagramby Penrose :Chaours.rv.pagesperso.orange.fr/
physics/Stars/Black/penrose.htm).

Our interpretation of Minkow's Diagram of space-time (DiagramofPenrose : Chaours.rv.pagesperso.orange.fr/ physics/Stars/Black/penrose.htm).

After Big-bang Fossil radiation or reappearance of the first light

Condensed universe

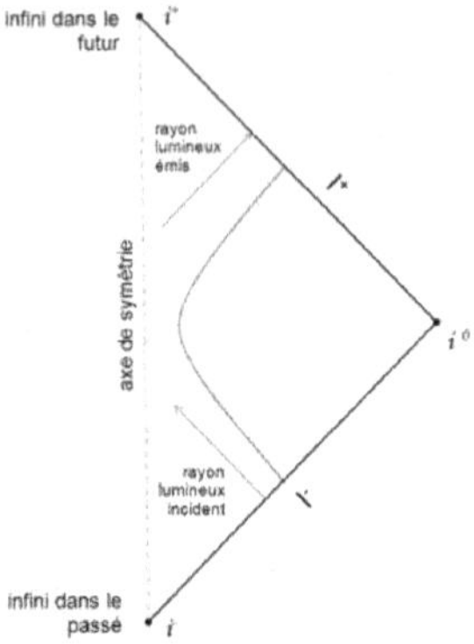

Before Big-bang first light (Ge 1 : 3 "God said there was light and there was light)

For us, this diagram is scientific proof that we do not think in a vacuum. This scientific diagram shows that there was indeed an incident ray of light before the Big-bang point (-) and that in the point io where the Universe totally condensed, nothing could be visible anymore, not even light and that the latter reappeared afterwards, at the point (+).

Section 4 : The Universal Cosmological Model versus other Pre-Big-bang theories

4.1. The theory of quantum nothingness

The theory of quantum nothingness is supported by Prof. Lawrence Krauss of Arizona State University. While claiming that there was nothing before the Big-bang, his theory is based on so-called virtual particles that are born and die in empty space in fractions of a second before the Big-bang.

This introduction is in line with the Universal Cosmological Model in that the Bible confirms that in the beginning ”the earth (Universe) was formless and empty”. Astrophysicists have shown that in the vacuum there are virtual particles.

Professor Krauss explained that if you remove virtual particles from a region of space, it will still have an energy density when it should not. It is from this nothingness that our Universe could have been born.

By analyzing Professor Krauss' theory of quantum nothingness, we discover that a lot of phenomena are missing between quantum nothingness and the Big-bang. These are indeed the phenomena described in the Bible at the moment when Moses speaks of ”a formless and empty land, until there was the separation of the waters from the waters” (Gen 1:1-6) or Big-bang.

Thus, everything that happened between quantum nothingness, a theory developed by Prof. Lawrence Krauss, and the Big-bang, was largely developed in Volume I. However, in this volume, we are going to use what we said in the previous volumes while revising it.

4.2. Cyclical universes

Cyclical universes are part of the theories supported by Albert Einstein in 1930 and later by Sir Roger Penrose, Professor at Bedford College London, Princeton University, and Wadham College Oxford. What is currently being debated is the model by Sir Roger Penrose and the Armenian mathematician Vahe Gurzadyan in 2010, which is called the Conformal Cyclic Cosmology (CCC) model. We explained the details in Volume II.

Indeed, in the cyclic model, a universe can have an origin and go through phases of inflation and expansion. This expansion goes up to a point and then stops, and after gravitation resumes this right. In this way, all galaxies contract again up to a certain maximum density, before the Universe rebounds again.

In this way, he will conclude that our Universe is just a continuation of the other universes that preceded us. One day, it will meet the same fate to give way to other universes again.

In this theory, Sir Roger Penrose doesn't know when the first universe began so that we can reach ours. Moreover, this first universe began by what process? It is through this we see the superiority of the biblical scheme of creation, as rendered in the Universal Cosmological Model, which offers a beginning for the Universe in quantum nothingness before we arrive at the Big-bang or "the separation of the waters from the waters".

In addition, the idea of a cyclic model has existed since the 1920s and Einstein supported it by proposing a variant in 1930. But in 1934, the American physicist Richard Tolman demonstrated that these cyclic models, including Einstein's, could not work because of the second principle of thermodynamics. Physicists know what this means. But generally speaking, at each stage, the amount of energy usable within it would only decrease. Thus, each expansion would be slower than the previous one, until a Big-crunch would occur without restart energy. (Cf. : Magazine " Les secrets de l'Espace ", Hors-série n°7, les mystères de l'Univers, what happened before the Big-bang ? p.11)

However, Sir Roger Penrose interpreted the diffuse cosmological background in the same way as in the Universal Cosmological Model. Indeed, for the Cosmological Model, the radiation of the Cosmological Background comes to us from before the Big Bang when it says in the Bible *"Then God said, let there be light and there was light"*. This light, we have said, did not appear because God spoke, but it was a way for Moses to demonstrate that it was part of the divine plan. This light appeared as a result of the contraction of the original particles in a movement described in the Bible as "a wind of God moving over the waters".

It is this contraction, which is often confused by supporters of cyclical universes, of a universe that contracted in a past life, as the Indians say.

For Penrose and Gurzadyan, the image of the cosmological diffuse background contains stains showing the presence of extremely weak concentric rings from earlier cosmic cycles. For the Universal Cosmological Model, this shows that there was a pre-Big Bang. As I have said, it is the astrophysicists themselves who will prove in their arguments the truthfulness of the Universal Cosmological Model and, consequently, the truthfulness of the biblical scheme of creation.

4.3. An infinite empty Universe

A Universe coming from an infinite vacuum is the theory supported by Prof. Gabriele Veneziano. It is the theory which is very close to the biblical story of creation through its evolution, except that the infinite, in the past, was characterized by a spiritual world, before a part of spiritual particles were transferred, in physical terms, into physical particles and became a formless and empty earth. I have spoken in physical terms because, in religious language, we will say that it was God who created this formless and empty earth signifying the Universe.

Coming back to the theory of an "infinite empty universe" supported by Prof. Gabriele Veneziano, the Universe has always existed and has remained almost empty during this time. Gravity finally took over and material began to agglomerate. This is in line with the biblical phrase "a wind of God moving over the waters" and this is what we have called cosmic deflation in the Universal Cosmological Model.

Again, according to the theory of an "infinite empty universe", the density of material is such that, in certain regions, it forms a black hole whose interior is undergoing a process of expansion. Here again, this comes back to the Bible through the verse "there was darkness on the surface of the abyss". The abyss is defined, according to the glossary of the Free Bible and Louis Segond, as a "mysterious world of the primitive ocean, a place of darkness and evil powers, of distress and perdition. "This definition is indeed consistent with the description of a black hole. If John Wheeler had known this definition, he could have kept the name "abyss" for black holes in 1967.

Indeed, the word abyss, used by Moses in the Bible in a literal way, fits with what the physicist John Wheeler explained in a scientific way regarding black holes as we can read :

"The term 'black hole' was coined by the American physicist John Wheeler in 1967 to describe a concentration of mass-energy that has gravitationally collapsed under its own gravitational pull and become so compact that even photons cannot escape this gravitational pull". (Cf. : Laurent SACCO,www.futurascience, Black Hole).

The last sentence of this quote says "that even photons cannot escape this gravitational force". This fits with the sentence "place of darkness and evil powers, of distress and perdition. "

Let us return to the theory of an "infinite empty universe" supported by Prof. Gabriele Veneziano who said : "As material continues to be swallowed by the black hole, its density increases to a limit imposed by physics".

This is where the second difference with the biblical story of creation came in. At this point, according to the Universal Cosmological Model, material has heated up to the point where the light (fires) described by the biblical phrase "God said that let there be light and there was light" is released.

Finally, the theory of "an infinite empty universe" supported by Prof. Gabriele Veneziano ended with "when it can no longer withstand quantum fluctuations, these cause material to expand outwards in a Big-bang scenario inside its black hole universe".

This is why I have quoted Prof. Gabriele Veneziano several times, saying that he offered a scheme close to the biblical story of creation. However, as we have just seen, the Universal Cosmological Model is more precise for understanding the Universe from before the Big-bang to the living.

4.4. Other theories supporting the pre-Big-bang era

We cannot analyses all the existing theories, but the common point for all of them is the existence of a black hole before the Big-bang. I myself was troubled by the word abyss in the Bible, in this stage of the Universe, and I was even more troubled when its definition referred me to the black hole. I almost stopped my research when I said : how can a black hole end up in the Pre-Big-bang? It was only after I understood what "space-time distortion" caused by a concentration of particles meant that courage came back to me.

Thus, in Volume I, I spoke of the two schools for the interpretation of the abyss, namely a Universe Pre-Big-bang with a black hole and a Universe Pre-Big-bang with a distortion of space-time interpreted by other theorists as being an abyss. And I have developed both theories in Volume I.

I discovered several articles testifying that the Big-bang came from a black hole. Even if I don't share what their authors say about what happened before, at least they confirm my scientific restitution of the biblical creation scheme according to which the abyss is assimilated to a black hole.

Among the other theories of the abyss as a black hole we have :

4.4.1. The Universe from another universe

– According to a hypothesis by three researchers from the Perimeter Institute of Canada, published in the journal "Scientic American", our Universe may have emerged from a black hole contained in a universe with a greater number of dimensions. They are Niayesh Afshordi, Associate Professor at the Perimeter Institute, Robert Mann, Affiliate Member of the Perimeter Institute and Professor at the University of Waterloo, and doctoral student Razieh Pourhasan. According to the same articles, their model seems to solve long-standing cosmological problems and - crucially - it produces verifiable predictions. The full comments that follow could be read in : https://ingeniumcanada.org/innovation/fr/recit/un-trou-noir-lorigine-de-lunivers.

– Professor Nikodem Poplawski, a physicist at the University of New Haven, Connecticut, has developed a theory that our Universe came from a black hole. He explains : 'Our universe should obey the same laws as the parent universe in which the black hole exists'.

He joins us when he continues : "I have published a paper that shows the consistency (of the black hole scenario) with Planck's observations of the cosmic background." We said, in Volume I, that the proof of the existence of the pre-Big- bang can only be read in the mirror of the cosmic microwave background or if sub-subatomic particles are ever discovered.

– Finally, for Professor Stephen Hawking, "Our Universe could be the interior of a black hole existing in another Universe. Black holes formed by stars and galaxies in such a Universe create new universes. A Universe can therefore give birth to billions of baby universes, which are formed by black holes. (Cf. : Magazine " the secrets of space#OUT OF SERIES N° 7 : What happened before the Big-bang ?).

The Universal Cosmological Model does not share its point of view on a multidimensional pre-Big-bang universe. If we were to go back in time, such universes would collapse and come back, as the American scientist Penrose states in his Model of Cyclical Universes and this last statement by Stephen Hawking.

4.4.2. Quantum field theory in curved space-time

This theory was authored by the physicist Leonard Emanuel Parker (born 1938), professor and former director of the Center for Gravitation and Cosmology at the University of Wisconsin-Milwaukee. In the late 1960s, Parker established a new field of physics - the theory of quantum fields in curved space-time. He discovered the physical mechanism known as cosmological particle production.

We have said that the pre-Big-bang era was characterized by cosmic deflation where the universe condensed, creating fluctuations in the vacuum. The post-Big-bang period, on the other hand, is characterized by inflation and expansion. Professor Parker suggested that the expansion of the universe, which for us has the same consequences as deflation, can create particles outside the vacuum. We read in an article : "His work has inspired research by hundreds of physicists and has been cited in more than 2,000 research papers ; it was credited in the memoirs of Soviet physicist Andrei Sakharov and helped Stephen Hawking discover the creation of particles by black holes". (Cfr : Leonard Parker ; Wikipedia).

A general prediction of this theory is that particles can be created by time- dependent gravitational fields (production of multigravity pairs), or by time- independent gravitational fields containing horizons.

Commenting on his theory, we read in another scientific article : "Some results indicate that the creation of particles from the quantum vacuum proposed by Parker is important for understanding how the particles that gave birth to our Universe were obtained from the inflaton vacuum". (Journey through the cosmos ; The vacuum and nothingness - edited by RBA Collecionables, S.A.U. Barcelona, 2020, P.144).

Conclusion on the Pre-Big-bang

Now, the Universal Cosmological Model gives scientists new proofs for thought for the pre-Big-bang cosmos. The Bible has spoken in understanding, and it is opportunity for scholars to extend it more. We noted the following major structures before Big Bang :

– The earth was formless and empty : Universe without form and empty at the start ; a physical nothingness

– The Darkness : seat of dark energy, dark energy, void energy. If we remove all the stars in the universe, there will only be a dark space containing the energy of the void.

– Surface of the abyss : contours of the black hole, area of entry for accretion of material.

– Abyss : black hole or first deformation of space-time ; place of concentration of the first original particles

– the spirit of God or a wind of God : Universal first force. Energy of material and on material, Pre-Big-Bang contractive force

– Waters : original materials at the start, sets of particles and materials before Big-bang which will evolve into materials after Big-bang.

– Lights : heat, fires, radiation

The combination of these seven sets resulted in the Big Bang or the separation of the waters from the waters (Ge 1 : 6). (volume I P.264)

Ge 1:6 *And God said, let there be a firmament between the waters, and let it divide the waters from the waters. 7 And God made the firmament, and divided the waters that were under the firmament from the waters that were above the firmament. And it was so. 8 And God called the expanse heaven. And there was evening and there was morning, the second day.*

We can schematize all that we have said about the Pre-Big-bang in the following way

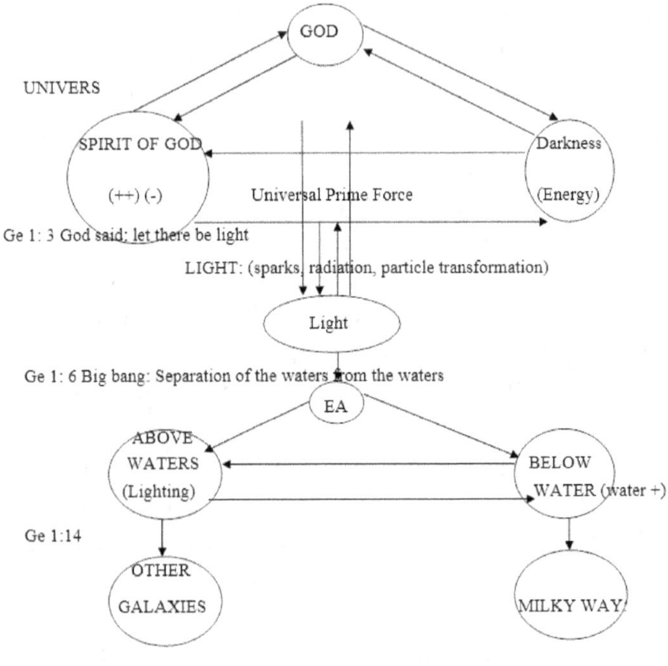

Ge 1:2 *The earth was formless and void, and darkness was over the face of the deep, and the Spirit of God was moving over the waters. 3 And God said, Let there be light. And there was light. 6 And God said, Let there be a firmament between the waters, and let it divide the waters from the waters. 7 And God made the firmament, and divided the waters that were under the firmament from the waters that were above the firmament. And it was so.9 And God said, Let the waters under the heaven be gathered together in one place, and let the dry land appear. And it was so.10 God called the dry land earth, and the heap of waters he called seas. 14 And God said,*

Let there be lights in the expanse of the heaven to divide the day from the night; and let them be for signs to mark the times and the days and the years;

The understanding of cosmology according to the biblical creation scheme, which scholars have neglected, believing it to be an act of faith, now becomes the goal to be reached in order to have true knowledge about the Universe.

If I have talked a lot about emptiness and it has been the object of my research to this day, reading books that drive me crazy, it is because it is a question that has always intrigued me, namely how from emptiness one arrives at reality according to processes and not spirituality.

The common point between the scientific pre-Big Bang theories and the Universal Cosmological Model is that the Universe came out of an abyss according to the Bible and a black hole according to these scholars. Apart from this common point, our understanding of the pre-Big-bang is different, but leads to the same result. For the Bible, there was a vacuum in the beginning. But for them, they don't know where the emptiness started, because they don't know the Principle of Creation of Rev. Dr. Sun Myung Moon.

Chapter IV :
The Big-bang

(Excerpt from Volume I and revised)

Ge 1:6 *And God said, let there be an expanse between the waters, and let it divide the waters from the waters. 7 And God made the expanse, and divided the waters that were under the expanse from the waters that were above the expanse. And it was so. 8 And God called the expanse heaven. And there was evening and there was morning : and it was the second day.*

Here is an illustration according to science of what happened from the Big Bang to the present day :

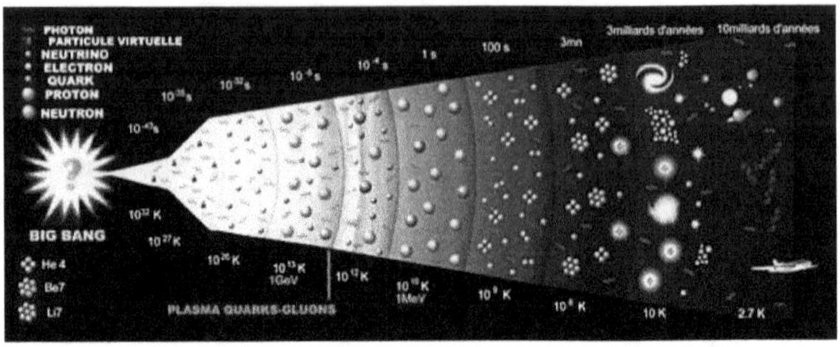

astropolis.fr/articles/history-of-the-world/the-zero-or-initial moment

The Big-bang, which we have also explained extensively in previous volumes of the series of books "The Verses of the Scholars", represents the zero moment when the Universe, strongly condensed with material called water by Moses, exploded. Nowadays in science, we no longer speak of explosion but of expansion.

In the biblical scheme of creation, this is represented by the phrase "the separation of the waters from the waters" (Gen 1:6) creating the expanse, demonstrating that there was something homogenous called waters by the Bible and which separated. The waters below serve the creation of non-bright spatial things, including the earth, and the waters above serving for the creation of luminaries, i.e. all those who emit light.

In the Universal Cosmological Model which is the scientific restitution of the biblical creation scheme, the Big-bang, which science has hitherto made the beginning of the Universe, was only a passage from a Universe in compression, but whose hatching or expansion spilled the resulting material.

Given the wide publication in these fields by astrophysicists, whose ideas we approve of, we do not want to go back over these notions after having discussed them at length in Volume I. The greatest discovery for me and the great revelation for everyone, which will correct more not only scientific books but also theological publications based on biblical commentaries, to learn that the Big Bang is biblical. Does this confirm that it is God who is the creator of the universe? It is up to each one to draw his or her conclusion after meditation; for this book does not impose anything by faith.

In this book we reserve our development on the Big-bang by answering, at the end, the questions of the internet users. However, the Big-bang has also created riddles that have not yet been solved, notably the question of dark material, dark energy and the force that activated it.

As the creator of a Model at a time when science has already taken off, I came very close to the origin of dark material in Volume I and spoke about the manifestation of dark energy, again in the second part of the same book, developing the thoughts of the Greek philosopher Empedocles (c. 493-433 BC), a disciple of Pythagoras and Heraclitus.

Concerning the third enigma, namely the force that activated the Big-bang, science speaks of the disassociation of the four fundamental forces. We have also added our point of view in the first two volumes of the series of books "The Verses of the Scholars", of which we introduce here some comments so that those who read this book can have an almost complete vision of the Universe.

Section 1 : The activation force of the Big-bang

We spoke about the biblical and philosophical interpretation of the activating force of the Big-bang in Volume I and the scientific interpretation in Volume II. Since this volume is read by those who only want to understand the universe in a general way, we put only the biblical and philosophical interpretation of the force that activated the Big-bang and let the scholars go to Volume II on the scientific interpretation. The three interpretations; biblical, philosophical and scientific allow us to find the answer to this concern of Capucine Casati who wrote an article with these terms :

"*What source can provide the amount of energy - called dark energy - necessary for acceleration ? This is where the most original theories come into play. None of them currently prevails*". (Cf. : Ciel&Espace magazine, August 2005, what is the expansion of the Universe ? by Capucine Casati).

1. Biblical and Philosophical Interpretations

I don't know if the philosopher Empedocles, who lived in the 5th century BC, knew the story of creation as told by Moses who lived ten centuries before him. And yet, it is these writings of Empedocles that allow me today to understand the scientific enigma on energy that made the Universe expand through the phenomenon called Big-bang.

For seven years, I had headaches because I could not assimilate my own words saying that it was from the action of darkness (dark energy) with the spirit (a wind) of God was hovering over the waters, that we ended up with the Big- bang. As both of these energies were present before the Big-bang, what bothered me was the way they acted.

The same applies to Yin and Yang. But the development of Empedocles's thinking helped me to understand how the events unfolded.

For Empedocles, or the ancient western cosmological model, we read this :

"*Originally, all the elements, united by Love, formed a homogeneous mass called the spherus. Discord penetrates it and separates the elements from each other».* (www.WIKIPEDIA : Empedocles).

In the Bible we read :

"*The earth was formless and empty; there was darkness over the surface of the deep, and the Spirit of God moved over the waters. God said, 'Let there be light. And there was light. And God saw that there was light good; and God separated the light from the darkness. And God called the light day, and the darkness he called night. And there was evening and there was morning : and that was the first day. And God said, Let there be a expanse between the waters, and let it divide the waters from the waters".* (Gen 1:1-6)

Still related to these quotes, for science, the Universe was condensed to the extreme before undergoing an exponential expansion called the Big-bang. However, the energy behind this expansion, according to science, remains a mystery. According to quantum mechanics, whose hypothesis is mostly accepted, after the Big-bang, it was gravity that broke apart in supergravity and this caused the expansion.

Returning to our quotations, Empedocles's explanation, inspired by God, allows me to understand all the models and the primordial force that was at work at the origin of the Universe. It can be summed up as follows : the unification of the elements, or the condensation of the Universe, was done by a single force called Love in the Western model, Yang in the Eastern models, a wind of God in the Bible.

For Empedocles, the discord, represented by the biblical darkness, although present and latent, only occurred when the spherus, or the biblical waters, or the condensed Universe, formed a homogenous mass. This is proved by the uniform temperature of the fossil radiation. This homogenous mass, or Condensed Universe according to science and waters according to the Bible, was condensed by the force of love according to Empedocles, by the spirit of God according to the Bible, by Yang according to Eastern philosophy and heated to the extreme. At this stage, the discord according to Empedocles, the darkness according to the Bible, the dark energy according to science, did not yet act because everything was governed by a deceleration defined by the Universal Cosmological Model of cosmic deflation.

1.1. The appearance of fossil radiation

But what caused these elements to lose their strength so that discord would enter into it according to Empedocles, so that gravity would disassociate according to science, and so that darkness would be put into action according to the Bible, and so that there would be Big-bang and inflation ? In fact, a major event has taken place, which only the Bible describes and which we reproduce in a scientific manner in the following paragraphs.

During its condensation by love, or the spirit of God, or the First Universal Force, the Universe emitted its first radiation defined in the Bible by the phrase : "God says : Let there be light ! And there was light". That is to say that the heat had reached a certain phenomenal degree. This is how astrophysicist B. Millikan, quoted in David Elbaz's lecture "In Search of the Invisible Universe", in his search for fossil radiation, spoke rather of the search for the "cry of God" which for him was the divine signature of the formation of atoms.

Moreover, the terms "cry of God" are justified because it was the first word of God expressed biblically through the phrase "God says : Let there be light...". "(Gen 1:3). Certainly, if B. Millikan was still alive, he would only have opened his mouth by my analysis, for he himself uttered this term while ignoring the first cry of God in the Bible.

It is this light emitted in the Universe that marked, according to the will of God for us believers, the countdown that allowed Hate, or darkness, or dark energy, to be activated in the Spherus, or in the biblical waters, or in the condensed Universe, to extend it again by a biblically defined phenomenon of the separation of the waters from the waters and the Big-bang according to science.

How do we draw this conclusion? At the emergence of this light, the Bible says "and God separated the light from the darkness". It is this radiation that science needs to study further, which has come to activate discord, or darkness, or dark energy, to act in a condensed Universe and ensure its expansion to this day. Hence their biblical name : "God called the light day and the darkness night" to demonstrate their eternity. Thus, we can say that the fossil radiation appeared before the Big-bang before being confined in the condensed Universe and reappeared after the Big-bang.

1.2. **Role of dark energy**

1.2.1. **From Pre-Big-bang to after Big-bang**

What role did dark energy play before the Big-bang, during the Big-bang and after the Big-bang?

Our research has shown us that current cosmology has not yet said anything about the source of dark energy, except that for science all energies originated in the Big Bang. All astrophysicists know for sure is that it is a form of energy associated with the vacuum of space, having a negative pressure in regions devoid of material.

Since human intelligence is complementary, I was the first to discover that it was Moses, through the Bible, and Empedocles, the Greek philosopher of the 5th century BC, who were the first to discover that dark energy is a pre-Big-bang force, and I might even add, that contributed to the Big-bang, not having withstood confinement.

According to Empedocles, when the Universe condensed under the impulse of love, or the spirit (a wind) of God according to the Bible, forming a mass called "Spherus", or the "waters" according to the Bible ; dark energy, or hate, or darkness, was already there according to the biblical phrase "There was darkness in the surface of the abyss". This contraction of the original material by the super-strength according to science, or a wind of God, or love according to Empedocles, heated the original elements until the first light from Pre-Big-bang according to the Bible came out.

Afterwards, hatred according to Empedocles, or darkness, or dark energy, caused discord, or an extent according to the Bible, or expansion according to science. And to this day, this expansion continues, according to science, under the impulse of the same energy called differently according to the times : darkness in the Bible, hatred in ancient times, by Empedocles, and black energy in present-day science.

This ancient philosophy is also in line with what some scientists suppose but do not dare to announce it clearly as in the commentary above :

"*very early in the cosmic chronology - probably before 10^{-35}s - the Universe would have experienced a dazzling expansion phase, inflation, when space, pushed by something similar to the current dark energy, would have expanded by a gigantic factor, say 10^{100}*". (Cf. : Ciel&Espace no.370, March 2001. Serge Jodra ; P.44)

So the dark energy, which Moses called darkness, as the first characteristic component of the endless vacuum is still preoccupying the scientific world today. With the speculation of great men of science, confirmed in the scientific documentary "New Theory of the Unified Field (Explanation of Gravitation)", in which they speak of dark energy as the primordial force in the creation of the Universe, we see that this notion is beginning to come closer and closer to the Bible without anyone, before my discoveries, having made the connection.

As for the physical properties of darkness that interest astrophysicists, I refer them to the same documentary to which I refer.

Another scientific explanation to understand dark energy is taken from this presentation :

"We know that the gravitational force acts like a magnet towards objects made of material. So it took another force to counteract this action. A force but this time, does not attract but repels. In other words, a repulsive force".

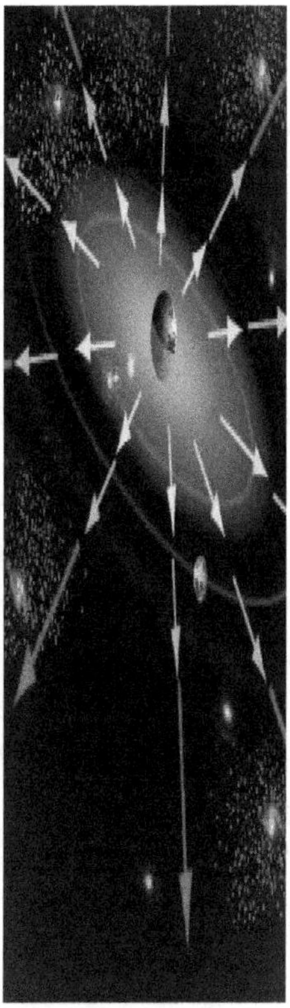

"Where does this repulsive force come from? While these effects can be observed, scientists are not aware of the composition of this force that counters gravitation. Hence its name : dark energy or dark energy. It is simply thought to be an energy from the vacuum and that it acts uniformly throughout the universe". (Cf.:lestoutespremieresfois.com/1er-coup-daccelerateur-de-lexpansion-cosmique ;
by Fabrice, Updated on 8 May 2014).

In these quotations, the gravitational force alludes to the spirit of God or a wind of God moving over the defined original material of the waters. Material is represented in the Bible by the word "waters". The repulsive force, called dark energy, is represented by darkness or vacuum energy. It was already presented at the very beginning of the pre-Big-bang Universe and this is affirmed by the biblical verse : "The earth was formless and empty, and darkness was over the surface of the deep, and the Spirit of God moved over the waters" (Gen 1:2). As we can see, everything that science has discovered as a great whole in the Universe is found in this verse.

Indeed, the author of the documentary [New Unified Field Theory (Explanation of Gravitation)] spoke of many interesting things about dark energy (darkness) as the source of everything. I can't include all these words in this book, so as not to enter into scientific polemics. Nevertheless, I invite you all to follow this documentary with great attention.

In conclusion, the biblical darkness at the beginning of the Universe was later described as a "night" full of energy; energy that was at first latent, then became active, contributing to the emergence of the Big-bang, until today. We have described this scientific process in Volume II.

Chapter V :
After the Big Bang

(Excerpt from Volume I and revised)

The post-Big-bang period represents the period after the zero instant of the Big-bang or "separation of the waters by the waters" according to the biblical terms. In the biblical scheme of creation, it began from the third day, whose paths we can follow :

Gen 1:2 "...the Spirit of God moved over the waters...". (Before Big-bang)

Gen 1:6 And God said, let there be an expanse between the waters, and let it divide the waters from the waters. (Big-bang)

– Ge 1:9 *"Out of the waters below* (1) *the dry* (2) (or earth) *appeared";* (beginning of the period after Big-bang).

Gen 1:13 And there was evening and there was morning, the third day.

– Gen 1:14 "*Out of the waters above* (1) *there were lights* (2)".

Gen 1:19 *And there was evening and there was morning, the fourth day.*

– Gen 1:20 *"Let the waters bring forth abundantly living*

creatures, and let birds fly over the earth toward the expanse of heaven.

Gen 1:23 *And there was evening and there was morning, the fifth day.*

– Gen 1:24" *God said, "Let the earth bring forth live animals after their kind, ...*

Gen 1:27 *God created man in his own image, in the image of God he created him ; male and female he created them.*

Gen 1:*31God saw all that he had made and behold, it was very good. And there was evening and there was morning, and it was the sixth day.*

We have developed all these periods in the preceding series of the books "The Verses of the Scholars".

According to the Principle of Creation, taken up in the Physics of Creation, the palpable material that followed this state began on earth with minerals, plants, animals and humans. This is consistent with the biblical scheme of creation. Some people, except astrophysicists, will tell me that in the Bible we talk about water. Indeed, water is dihydrogen and oxygen, as we can read ; "In astrophysics, the term metal refers to any element heavier than hydrogen or helium". (Cf. : Revue

La Recherche No. 551, P.34, 09/2019, edited in Paris, Une nouvelle source d'information sur le cosmos, S.G.).

These are the minerals which moreover began by forming the first stars.

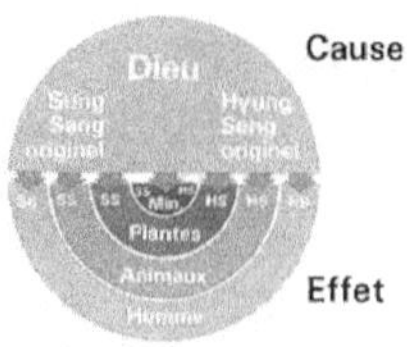

Source : Divine Principle Level 4, p18

All the scientific publications in natural sciences, anthropology, etc., contribute to enriching the post-Big-bang era and we do not see what we can add apart from demonstrating that the Bible, or the biblical scheme of creation, does not deviate from the results that are being discovered by science. This is what we are going to work on in this chapter by annexing some excerpts from Volume I where we have touched on some very important points that caused the biblical scheme of creation to be rejected by science because of recent scientific discoveries.

I know that many people expect me to tell them how things happened from the third biblical day, which marked the beginning of the post-Big-bang era, to the formation of man, but they only have to refer to Volume I.

Section 1 : The stages of the evolution of the universe according to the Bible

Gen 1.1. *God said, God made, God created and the spontaneous breaking of symmetry*

By symmetry, we said, we imply a conservation of properties, a kind of global stability even within a movement. The rupture of symmetry therefore supposes that a system where something that was preserved suddenly stops doing so and gives rise to an unexpected phenomenon.

The history of creation, for science, is about spontaneous breaks in symmetry. Moses, in the Bible, divided them into six days, which are part of the six periods of creation. However, within each period there are unexplained phenomena. For example, if we take the first day, or the period minus one in science (i.e. not yet recognized), we can ask ourselves the following question :

In the verse "God created the heavens and the earth (universe)" ; how did the physical (earth) appear from the spiritual ? For science, there has been a break in symmetry somewhere. The symmetry was a stable, self-preserving sky. And as a result, a part of the sphere of the heavens, filled with spiritual particles, was inexplicably transformed into physical particles and anti-particles in "a formless and empty earth".

So a break in symmetry implies a brutal change in the laws ; a transition, physicists would say, a revolution for all those who study social change. For science, many other properties of material, all in fact, have arisen within such transitions that mark the history of the Universe.

All this is to say that scientists are aware of the

miraculous phenomena "God said, God made, God created" by observing phenomena with missing links in the Universe. They have replaced this with the term "break in symmetry", which means the sudden, even spontaneous passage through a state in which a new parameter appears, different from what should have gone before.

This is confirmed by the quotation below :

"Discoveries in physics over the last few decades have led us to attach great importance to the concept of broken symmetry. The evolution of the universe since its birth is seen as a succession of broken symmetries. When it emerges from the Big Bang, the universe is symmetrical and without structure. As it cools down, it breaks each symmetry after another, thus allowing the appearance of an increasingly differentiated structure. The phenomenon of life itself naturally takes its place in this painting. Life too is a breaking of symmetry."

(Cf. : F. J. Dyson, Infinite in all Directions, Harper and Row, 1988).

As far as creation is concerned, until now scholars have observed discontinuous phenomena, giving way to other phenomena. However, they have not yet found a causal effect. In the biblical scheme of creation, the spontaneous breaking of symmetry took place in the places where Moses said "God said, said made or God created".

If science itself cannot find answers to these phenomena, in his time Moses would have attributed it to whom as a prophet and scholar ? To God, of course. But scientists prefer to speak of nature's chance. They mock the believers when they talk about "God said, God made, God created", but they want the believers to believe them when they talk about the spontaneous breaking of symmetries.

For a more precise example of the breaking of symmetry in the animal kingdom, it is man and monkey. Indeed, the monkey is often scientifically considered to be the ancestor of man. However, what we can see is that there was a break in symmetry which caused us to pass from the shape of a monkey to a being with a shape which allows him to do anything that goes through his head.

Biblically, after the creation of the animals, God put a pause. To finally say : "let us create man in our image". We see directly that what is going to be created or formed again will be different from the previous one. Between the previous and what follows, scientific analyses have revealed that there is a break in symmetry, because the missing link is nowhere to be found. This break in symmetry is sometimes described in biology as a mutational accident of the genes.

2. The biblical scheme of creation

Ge 1:2 "*...the Spirit of God moved over the waters...*". (Pre-Big-bang)

Ge 1:6 *And God said, Let there be an expanse between the waters, and let it divide the waters from the waters.* (Big-bang)

Ge 1:9 "*Out of the waters below* (1) *the dry* (2) (or earth) *appeared*"; (beginning of the period after Big-bang).

Gen 1:13 *And there was evening and there was morning, the third day.*

We have already explained that the phrase "God said" is to demonstrate that this was part of his will. But evolution has followed a scientific pattern characterized by what science calls breaks in symmetry.

All these phenomena, or evolutions, take place on the same original material, which is called, by default, "water" according to the Bible, be it Pre-Big- bang, during Big-bang and after Big-bang.

1.2.1 Illustrative table on biblical and scientific cosmology

BIBLICAL PERIODS	VERSES	CREATION	KEY WORDS	SCIENCE
SPACE-TIME	Ge 1 :1	In the beginning, God created the heavens and the earth.	-Heavens (Spiritual World),- Earth (Physical world)	Before the Big-bang
DAY 1	Ge 1 :2-3	2 The earth was formless and empty; there was darkness on the face of the deep,and the Spirit of God was moving over the water. 3 God said: "Let there be light ! And the light appeared"	- Empty – Darkness – Abyss -Spirit of God –Waters - Lights	
DAY 2	Ge 1 :6	God said, "Let there be an expanse between the waters, and let it divide the waters from the waters."	- Extent-Water separation	The BIG-BANG
DAY 3	Ge 1 :9-11	9 God commanded, "Let there be a dome to divide thewater which are below the heavens be gathered together in one place, and dryness appear. And so it was.10 God called the dry land earth, and he called the cluster of waters seas. "11 Then God said," Let the earth bring forth greenery "	– Seas – Earth – Herbs (Moleculesof life)	- Formation of elements up to galaxies (stars, lights and planets)
DAY 4	Ge 1 :14	14 God said: "Let there be lights in the expanse of the sky to separate the day from the night; let them be signs to mark the epochs, the days and the years ; "	-luminaries (Stars)	
DAY 5	Ge 1 :20 Ge 2 : 5-6	20 God said, "Let the waterbring forth living animals in abundance, and let birds fly over the earth to the expanse of heaven."	-marine beings (Plants) – seabirds (plants, bacteria, Pisces) – flying beings	-marine beings (Plants) – seabirds (plants, bacteria, Pisces) – flying beings
DAY 6	Ge 1 :24 – 27 Ge 2 :7 Ge 2 :19 Ge 2 :21 – 22	24 God recommended, "Let the earth produce allkinds of living animals , and domestic, and wild, large and small." 26 Then God said: "Now we will make human beings; they will be like us and resemble us. 27 He created them male and female "	– – Terrestrial animals male	– – Terrestrial animals male

Source : Compilation

1.2.2. Conception of the word "land" before and after the Big-bang

We have said that before the Big Bang, the word 'earth' was used in the Bible to refer to the universe and this is through the verse 'the earth was formless and empty' (Gen 1:2). And we said that something that has no form and is empty is like a nothingness which was none other than the universe at its beginning. It is in this emptiness that several phenomena took place that we are taught in quantum mechanics, which is part of the physics of creation.

For the Universal Cosmological Model, the universe began in a vacuum until cosmic deflation condensed all energy into one point. Then there was Big- bang, followed by cosmic inflation.

Because of the cosmic inflation and the extreme heat just after the Big- bang, the different laws and theories of quantum mechanics cannot be applied, according to the Universal Cosmological Model. For the Universal Cosmological Model, these different laws and theories can only be applied in an empty universe. Therefore, even if their authors, or creators, do not agree, several laws and theories of quantum mechanics concern the pre-Big-bang period, especially those related to the appearance of real particles from the vacuum, that is to say from the moment when the Bible spoke of a formless and empty earth (Ge 1:2).

In fact, cosmic inflation was initially characterized by the repulsive force of particles, which resulted in the expansion of the Universe, before the latter cooled down and the quarks came together to form the first elements, according to the current scientific conception.

This is why the laws and theories of quantum mechanics do not agree with the theory of general relativity that explains the Universe after the Big Bang. However, string theory, which is physically close to the biblical scheme of creation, tries to unite them, but will only succeed if they 'fit' into the Universal Cosmological Model.

In the post-Big Bang period, according to the Bible, we are still told about the earth, not as formless and empty (Ge 1:2), but as a dry mass, "God called the dry land earth, and he called the heap of waters seas" (Ge 1:10). Here we see that the material has given way to the immaterial and we are in the post-Big-bang period. Let us read :

Ge 1:2 *The earth was formless and empty : there was darkness over the surface of the deep, and the spirit of God moved over the waters.* (Pre-Big-bang).

Ge 1:6 *And God said, let there be an expanse between the waters, and let it divide the waters from the waters. 7 And God made the expanse, and divided the waters that were under the expanse from the waters that were above the expanse.* (Big-bang)

Ge 1:9 *And God said, Let the waters that are under heaven be gathered together into one place, and let them appear. And it was so.* (After Big-bang).

Ge 10 God *called the dry land earth, and he called the heap of waters seas. God saw that it was good.*

The key words here are the waters in one block, the separation (expanse and expansion) of the waters into two blocks, the formation of the earth from the waters below the expanse.

Why, after the Big-bang, did Moses speak directly about the planet earth ? Because, it was the only one who could interest his contemporaries. The other galaxies appeared as stars and were referred to as luminaries (Ge 1 : 14).

1.2.3. Cosmological parallelism between the Bible and science

ACCORDING TO THE BIBLE			ACCORDIN G TO SCIENCE (COMPILATIONS)		
CYCLE	DAY	CREATION	**CYCLE**	ERA	APPEARANCE
Pre-Big-bang Ge 1 :2	1st	- darkness (dark energy - abyss (black hole) - Spirit of God (energies) - waters (original particles) - light (lights, heat, rays))	Before the Big- bang	- aporia	Gray area for science. No scientist dares to express a point of view for fear of being ridiculed one day. But, several assumptions are starting to hold true with Ge 1: 2 including string theory and others that imagine the Big-Bang Before.
Big-bang Ge 1 :6	2e	- Separation of the waters from the waters	1st	-15,000,000,000 years	Big bang (explosion or expansion of the universe) followed by the formation of the elements.
AfterBig-bang Ge 1 :9	3rd	Material Universe: dust, planetesimals, gas,life molecules	2nd	-14,999,700,000 years at – 14,900,000,000 years	Formations of helium atoms, hydrogen atoms, cold clouds of cosmic gas, galaxies
	4th	formation of stars and planets	3rd	-10,000,000,000 years	formation of stars and planets

Ge 1 :14					
Ge 2 :5 Ge 1 :20	5th	First plants Seaand flyin g animals	4th – paleozoic	-4,100,000,000 years to – 400,000,000 years	– 1st chain of nucleotides, – shallow oceans, carbon-based compounds, – first bacterial cells, DNA, – complete development of bacteria, the oxygen level in the atmosphere reaches its current level: 21% – first plants (Paleo-zoic) —First land animals (insects, amphibians)
Ge 1 :24	6th	Mamm and als humans	5th Mesozo ic-Jurassic	- years 200,000,00 0 to – 4,000,000:	- flying reptiles, 1st mammals (Mesozoic) - flowering plants, first primates (Juras-sic) - 1st man (Australopithecus)

Source : compilations

Section 2 : The molecules of life

2.1. Chronological contrast between the 3rd biblical era and the 2nd scientific era

As said before, in what is devoted to the post-Big-bang era, we only address the points that made scientists reject the Bible. After explaining the difference between the word "earth" at the beginning of the Universe (Ge 1:1-2) and on the 3rd day of creation, many scientists point out the contrasts between the 3rd biblical era, called the 3rd day, and the 2nd scientific era. They raise this point to say that the biblical scheme of creation is not real, because how plant life could have appeared before the luminaries.

Indeed, in the table above, there is a discrepancy between the 3rd day of creation and the 2nd cycle of science regarding the formation of the earth and the appearance of plants before photosynthesis, but this is a misunderstanding.

Let us first read this verse :

Ge 1:12-13" *And the earth brought forth greenery, grass yielding seed after its kind, and trees yielding fruit and having seed in them after their kind. And God saw that it was good. And there was evening and there was morning, and it was the third day.*

We have said that Moses spoke of the initial stage and the final stage (then there was evening and there was morning). Due to the fact that after the Big-bang the galaxies began to form (as science attests in the second stage), the earth, being a galactic element, was already included according to God's purpose.

Describing the germination of plants at this stage, it alluded to the creation of life molecules at this early stage in the history of the Universe. This means that the same molecules can also be found in other stars, but in a static way, while those on planet earth will bear fruit. Thus, today, the existence of organic elements in certain stars is beginning to complicate some of the theories of exobiologists, especially elements that are today at the basis of the composition of living beings.

As we have said, Moses had already designated these elements on the third day of creation by molecules which he described as plants in Ge 1:11-12, when we were still in the primitive stage of creation. Even in the Ancient Oriental Models, which we mentioned in Volume I, these elements were referred to as wood. But the vegetable kingdom, like all living kingdoms, began to appear from the fifth day of creation.

We said at the beginning that if the first chapter of Genesis is very religious, where Moses attributes all creation to God, the second chapter, on the other hand, takes us into scientific cosmology. And in this chapter we can clearly see that plants appeared in the fifth stage, thus joining science as we can read :

Gen 2:5 "*When the LORD God made the earth and the heavens, there was not yet any shrub of the field upon the earth, nor any herb of the field sprouting up : for the LORD God had not caused it to rain upon the earth, neither was there any man to till the ground*".

2.2. The appearance of the molecules of life

In the scientific documentary "The origins of the earth" we are told that at the origins of the earth, the earth's crust was already carrying life when it was still undergoing all kinds of bombardments. Even before the moon separated from the earth, according to the same documentary, there were already the molecules of life. Hence the projection that Moses made of the earth. It is a documentary that I recommend to scholars who also evoke as a mystery the origin of water, for which we offered our clarifications in Volume I.

Another comment comes from this study by the scientists :

"*Water appeared on Earth 4.35 billion years ago. This finding by NASA researchers is based on the analysis of zircon crystals extracted from a rock in Jack Hills, Australia... Thus, a condition essential for the appearance of life already existed only 200 million years after the formation of the Earth*" (Cf. : Ciel&Espace Août 2005 ; De l'eau liquide très tôt sur terre, p.26).

In fact, in the Bible, in speaking of the creation of the earth on the third day, we have said, Moses also speaks of the appearance of plants, whereas according to science, plants appeared at a time corresponding to the 5th biblical day. We have said that they were molecules that brought life to this planet, as we can read :

Ge 1:9 " *God said, Let the water under the sky be gathered into one place, and let the dry land appear." And it was so. 10 God called the dry land "earth," and the gathering of the water he called "seas. And God saw that it was good. 11Then God said, Let the earth produce vegetation : seed-bearing plants and fruit trees on the earth bearing fruit with seed in it according to their kinds. And it was so... 13 And there was evening and there was morning : the third day.* "

Indeed, these verses seem to be in contradiction with science as far as the appearance of plants in the 2nd cosmological cycle or 3rd day according to the Bible is concerned. Going to chapter 2 of Genesis ; Moses spoke of the appearance of plants on the 5th day, which corresponds with science. Is there a contradiction between these two verses? No, today even the scientists know that in the supernovae that came at the end of the life of the stars, there is the liberation of the atoms that formed inside, including the molecules of life.

This is how Moses, already knowing the finality of what will happen to the viscous mass that will become the earth, speaks of the plants on the third day, which were nothing more than molecules that will evolve to form life, and the latter began with an underwater being, baptised LUCA by biologists, which in its evolution subdivided into bacteria, eukaryotes and archaea. It is the eukaryote, (LECA, Last Eukaryotic Common Ancestor) which is considered by biologists to be our last common ancestor, starting with an underwater flora, passing through plants, fungi and animals (Cf. : The Challenges of Science : The Origin of Life, edited by RBA coleccionables, S.A.U. 2018, Barcelona, Pp.105-124).

Moses in this verse already unveils God's plan for planet Earth. As for the creation of the plants themselves, this took place on the fifth day as mentioned in the second chapter of Genesis.

What confirms the idea that life-bearing molecules were already present during the period of galactic collisions is the discovery of organic molecules in comets and some planets.

If in some planets, comets or stars, these materials remained static, on the earth, there was evolution until life, and this, as we have said, as have scientists, because of the privileged position that the earth occupies in relation to other stars and the conditions, some have not yet been discovered.

Let's read for this the comments taken from the internet on the organic molecules of comet 67P/Churyumov-Gerasimenko after the analysis of

Philae's data by the researchers of the CNRS of France, published in the edition of 31 July 2015 in the journal Science :

"*Since the materials of comets have been very little modified since their origins, this means that in the early days of the Solar System, organic compounds were already agglomerated in the form of grains and not just as small molecules trapped in ice as previously thought. It is such grains that, when introduced into the planetary oceans, could have favored the emergence of life there*".

This is a scientific idea that is in line with what we are saying.

Until now, the emergence of this life has only been found on planet earth, even if we are now beginning to speculate on certain stars.

What we will have to remember in this section is that after the Big-bang, there was also the formation of organic elements in the first groupings of material. This is what makes it possible to find them everywhere.

2.3. Process of the appearance of life molecules

What we can add is already summarized in this table :

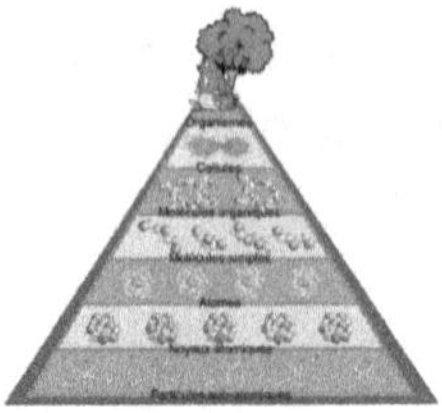

astrocanada.ca/the_contemporary_cosmos_material/Fra

Indeed, after the Big-bang and the other processes that followed, the organic molecules were redistributed throughout the Universe. What was on planet earth, with the environmental combination created by its atmosphere and other as yet unknown processes, should give rise to the life that Moses described. In Volume I, we talked about how one of these processes concerning the appearance of plants took place in the Galapagos Islands.

Section 3 : The solar system

3.1. Formation of the solar system

Ge 1:9" *God said, "Let the water under the sky be gathered into one place, and let the dry land appear. And it was so. 10 God called the dry land "earth," and the gathering of the water he called "seas.". And God saw that it was good. 13 And there was evening and there was morning : the third day.* "

Ge 1 : 14 *And God said, Let there be lights in the expanse of heaven to divide the day from the night; and let them be for signs to mark times and days and years; 15 and let them be for lights in the expanse of heaven to give light upon the earth. And so it was. 19 And there was evening and there was morning, and it was the fourth day.*

Comparing these two verses with the chronology of the Universe, I really shouted Eureka, like Archimedes. Not only was the discovery of the Big-bang in the Bible the greatest discovery for me, but I was also completely unaware that the earth could be older than the Sun.

This effectively makes the Biblical Model, reproduced in the Universal Cosmological Model, a scientific model. Who in the time of Moses, looking up into the physical sky, could have the idea that the stars, the Sun and the Moon in the solar system had reached maturity after the Earth? Why am I talking about maturity? Because the materials that are part of the constitution of these stars, called "waters" by the Bible, were all projected at the same time into space, a phenomenon defined by the biblical phrase "from the waters below, the earth emerged, and from the waters above, God made the lights. "

Let us limit our analysis to the case where Moses only spoke of the solar system, the only one that was accessible to his imagination.

For the earth, we are told of the third day, or third era, and for the lights we are told of the fourth day, or fourth era. What was my surprise when I read these scientific texts?

"*The age of the Earth's oldest rocks is measured in the laboratory by dating methods based on the decay of radioactive elements. The age of the Earth thus obtained is (4.5 - 0.1) billion years. Lunar samples lead to values of the same order of magnitude. The analysis of meteorites, which are known to be the oldest bodies in the solar system, allows a measurement more directly related to the age of the formation of the solar system. The study of radioactive isotopes in meteorites, such as uranium, lead and strontium, for example, leads to an age of (4.55 - 0.05) billion years. The Sun, not having at that time begun to burn its hydrogen into helium, is a little younger. An analysis of its evolution since the condensation phase of the solar system leads us to attribute an age of (4.49-0.05)billionyears".*(Cf.:https://www-n.oca.eu/vigouroux/chp1/chp123.html ;Âge du Soleil, Vigouroux Anne, Friday 13 September 1996)

As can be seen, all the stars of the solar system have almost similar ages confirming the biblical story of the creation of the third day and the fourth day. The Bible does not say that the Sun was created on the fourth day, and that the maturity of the earth preceded that of the Sun and the moon.

First of all, according to science, the Sun, the earth and the other stars of the solar system were all projected by the same mass, thus confirming the verses that spoke of the waters below and the waters above coming from the same body of water (Ge 1:6-7). Thus we can no longer say that the Sun is older than the Earth or vice versa.

Again according to science, astrophysicists have all the sources, of the mass that was projected, 99% of the material agglomerated to form the Sun and the 1%, called the waters below by the Bible, was used for the formation of the other planets, including the Earth.

If, for the scientists, it was the accretion of material that was important to speak of the Sun or the Earth, for God and Moses it was the planets, the grains of dust of what would become the Earth.

These planetesimals were called the debris collected in the disc. They were dust which, having gathered, ended up forming several coherent objects. At the time of their projection, what would become the planet Earth was already located, detected in its orbit and described as life-bearing.

I know that my readers are very curious to understand how things happened from the moment the Bible speaks of the separation of the waters, especially the waters above and the waters below, until they formed stars. The Belgian scientist, Father Georges Lemaitre, and the English scientist Stephen Hawking, predicted that one day there would be a cosmological model accessible to all and that was part of their concerns. This is what we are in the process of accomplishing.

3.1.1. Biblical water is the precursor to hydrogen

First of all, when Moses talked about water, he was not talking about our running water. This was in fact the name he gave to the particles that began their cycle in the pre-Big-bang, through the verse *"the spirit of God moved upon the waters"*, passing through the furnace of the Big-bang, known as "the separation of waters from waters" and going on to form the structures of our present

Universe by the waters above and the waters below. That is to say that all the structures of the Universe should have a common element related to their beginning.

This common element was discovered by Henry Cavendish in 1766, in the 18th century, and Lavoisier nicknamed it hydrogen, unaware that he had just joined the Bible.

In fact, when he decomposed water, Lavoisier called one of the elements hydrogen; (Hydro = water) and "gene" (engendered). This means what generates water. According to various scientific sources, this element occupies 85% of the visible Universe.

Moreover, not only is hydrogen one of the essential constituents of water,

i.e. one of the molecules vital for life, it is also the most abundant, constituting stars like molecular clouds. So Moses was indeed right to use the term water, which does not necessarily represent liquid water, but original material.

All astrophysicists believe that hydrogen is one of the elements that form outside the stars. In fact, it is the element that is the basis for the creation of Everything. Moses could only use the term "water", and in the end Lavoisier confirmed this with the word "hydrogen" when these particles formed the first nucleus of the Universe.

The processes of the formation of what the Bible calls luminaries (Sun and stars) began with the fusion of the particles until they formed hydrogen at the beginning of the Universe. This, on heating, eventually decomposed into helium. After the fusion of helium and hydrogen, carbon was formed. The carbon produced in this way fuses with helium, creating oxygen and so on. The repetition of this type of process - depletion of the nuclear fuel, followed by collapse and heating - creates the x's of the various chemical elements known down to iron.

All these created elements, are found in the form of fluids and plasmas and are maintained by gravitation under a ball that we call the star. Iron is generally considered to be the final stage in the chain of nuclear fusion reactions within stars. Thus, at this stage, fusion cannot continue to release its energy by building heavier elements. The ball then begins to absorb its energy instead of emitting it and this causes it to implode. Once the star, or ball, explodes, the increase in temperature it gives off creates even heavier elements.

By exploding, through a phenomenon called supernova, its outer layers propel into space elements they have made inside them and the gases we find in nature, including molecules of life. This is how, according to science, all the chemical elements we find in nature, as classified in Mendeleev's painting, are formed. It all starts with hydrogen, which means what generates water, and Moses had already baptized these subatomic particles with this name.

And what about the formation of the non-shining stars called planets, by the waters below according to the Bible?

If luminaries such as the Sun capture most of the material to be formed by fusion, the other debris, which finds itself in space far from any fusion process to initiate thermonuclear reactions, will also begin to assemble by the force of gravity in the nebular disc.

This nebular disc, for its part, cools, becomes thinner and narrower, taking on an annular structure where a condensation of denser corpuscles (dust, crystals, ice, rocky bodies, etc.) appears, in a wide variety of orbits that will eventually form the debris called planetesimals. This dust which gathered together, ends up forming several coherent objects in their projection orbits, forming planets, including the Earth. This is why in each planet we find the same elements as those that make up the stars, confirming the Big-bang or separation of water from water.

Moreover, these planetesimals agglomerate faster than the appearance of the luminaries. However, it is the latter that will come to shape them into planets, asteroids or comets by their bombardment of heat. By this you mean that there are gaseous or telluric planets that sometimes escape this rule.

3.1.2. The Earth and the Sun have the same chronological period

First of all, as I said, the masses of what will become of the Earth and the Sun were all projected at the same time. Moses referred to them as the waters above and the waters below.

Moses did not say that God began to create the light fixtures on the fourth day, but he said "from the waters above God created the light fixtures". As can be seen, the waters below and above were all projected on the third day. Thus, the third day and the fourth day are linked and there is no discrepancy. Whether science says that the Sun is older than the Earth or investment, has no impact on the Biblical 3rd and 4th days because their seeds were projected at the same time.

The same seeds planted in different places and under different conditions will also grow differently. This is how the Suns and stars remained in fusion, the earth and some planets became telluric and the other planets remained in the primitive gas stage. Those that have not yet begun their formation are still in the nebula stage, i.e. the gathering of dust and hydrogen gas in interstellar space. Thus, the ninth planet in the solar system could be a nebula hidden in darkness.

Moreover, talking about the creation of the Earth on the third day by the cluster of waters below the expanse, and of luminaries on the fourth day, this amounts to saying that our oceans, from which the Earth originated, must have been older than the Sun. This truth troubled me a lot until I came across this article on leMonde.fr published on the Internet in m blogs on 26/09/2014 by Guillaume Cannat :

"By studying the origin of the water in the Solar System, researchers have just concluded that a large part of it would have formed before the birth of the Sun and the Earth".

This explains why science sometimes presents us with the image of a rocky earth covered with magma when it was formed, and sometimes with an earth covered with ice because the sun's rays were not yet ready to melt it. In this case we can say that this first icy ball, which reflects the Biblical Genesis, was indeed older than the luminaries or the waters above, which had not yet acquired their stability.

Moreover, Moses, the scholar of scholars raised these points in the second chapter of Genesis by the following verses :

"When the LORD God made the earth and the heavens, no shrub of the field was yet upon the earth, neither was there any herb of the field yet sprouting; for the LORD God had not rained upon the earth, neither was there any man to till the ground". (Gen 2:5).

This was only in the period that could be described as the Ice Age, the chronology which historians have only to modify, if this does not correspond to our comments. I am not the only one to reason like this, let's read what the scientists have to say :

"In short, over the 4.56 billion years that elapsed since the formation of the earth, the history of the first 500 million years is very difficult to reconstruct. However, an

American-Australian team has just published astonishing results : they not only affirm that the first continents were already in place more than 4.4 billion years ago, but that in addition to that, water was flowing on their surface... This supposes that the very young planet cooled down at record speed, in less than a hundred million years, and moreover obliges the geophysicists to review their model of earth formation! "(Cf. : Ciel&Espace n°370. March 2001. P.18 by Azar Khalatbari).

If we meditate very well on this verse in relation to all these scientific commentaries; "*God says : Let the waters below the Heaven be gathered together in one place, and let the dry land appear. And it was so*" (Gen 1:9), we can say that the earth was freed from the ice when it melted. This is how the planet Earth is made up of 70% water and only 30% of the dry part.

We can still read this other comment :

"*Water appeared on Earth 4.35 billion years ago. This finding by NASA researchers is based on the analysis of zircon crystals extracted from a rock in Jack Hills, Australia*". (Cf : Ciel&Espace n° 423 - August 2005; Liquid water very early on Earth, p.26).

For the Universal Cosmological Model as we argued in Volume I, the first continent sprang up in the same way as the Galapagos and other islands in the oceans, melting the ice that covered it. Our proof, the transformation of the planetesimals resulted in the rock containing the zircons whose studies were used to date the Earth. If the ice covered the surface, seismic movements in the subsoil, making the water liquid, were already transforming these granites into rock.

This being the case, all the scientific comments about the appearance of water on the Earth must be re-examined, as they were previously based on the view that there was the rock and that the water came to us from meteorites.

Below is a diagram of the solar system today : the proto-sun became the sun star and the planets formed around the sun in the protoplanetary disc.

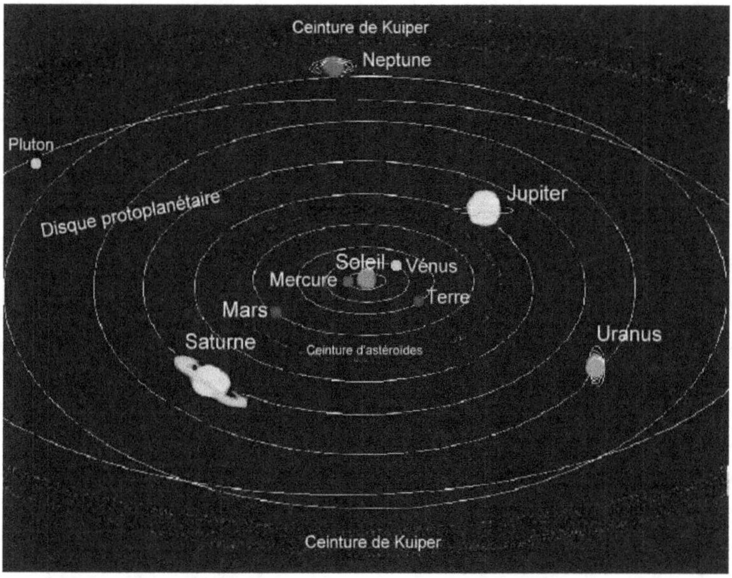

Image of the current solar system : the sun and its protoplanetary disc made up of planets and asteroid belts.

vinsvignesvignerons.com/Geology/Geologie-de-la-France

All the above planets were projected by the same mass but experienced different growths.

Section 4 : The plant kingdom, aquatic and flying beings

(5th biblical day)

4.1. The plant kingdom

4.1.1. the third day's prediction

Let's go back again to the third day of creation where we said that some people reject the Bible because it mentions the germination of plants before the Sun was formed.

Gen 1:11" *Then God said, " Let the earth produce vegetation : seed-bearing plants and fruit trees on the earth bearing fruit with seed in it according to their kinds. And so it was. 13 And there was evening and there was morning : the third day. »*

By following these thoughts, we discover that everything starts from the infinitely small to the infinitely large : greenery - grass - fruit trees. We see that the whole process is detailed in the Bible in a logical way. Scientifically, the process of the appearance of life is very complex. But the most important point, even in science, is that in creation everything starts from the infinitely small to the infinitely large.

We have already said that the third day of creation corresponds to the second era of science and that it was a period of galactic formation and collisions, and without life. Moses spoke of plants on the planet earth to express the presence of life molecules in this dust heap, as described in the scientific documentary "The Origins of the Earth"; written by Alice Harper and Nigel Henbest.

4.1.2. Life appeared on the fifth biblical day

However, going back to the second chapter of Genesis, we will see that life did not begin until the fifth day even though the pattern was described on the third day. Let us read :

Gen 2:4 " When the LORD God made the earth and the heavens";

Gen 2:5 "...and no herb of the field yet sprang forth : for the LORD God had not caused it to rain upon the earth...".

Ge 2:6 'But there arose a vapour from the earth, and watered the whole face of the ground'.

In Ge 2:5 we are told that no grass sprouted, because there was no rain yet. Then, in verse 6, we are told that the earth was watered. This shows that it was after the fourth biblical day when God created lights (Gen 1:14-19). This means that the first plant appeared in the 5th stage, as science shows in the table at the beginning. This is confirmed by the following verse :

Gen 2:7 "*The Lord God formed man out of the dust of the earth".*

Man was created in the sixth stage. Therefore, plants, according to the above verses, were formed in the fifth stage. But for Moses the formation of earth and life formed a whole and this began with the unicellular algae which he could only call plants. Moreover, according to different scientific sources, these algae (cyanophyceae) sowed, some of them into plants, phytoplankton, and others into marine beings, zooplankton.

Moreover, when God created man, he said to him :

"Behold, I give you every herb bearing seed that is on the

face of all the earth, and every tree that has fruit of the tree and bears seed, to you it shall be for food". 30 And to every beast of the earth, and to every bird of the air, and to everything that moves on the earth, having the breath of life in itself, I give every green herb for food" (Gen 1:28).

This proves that the creation of plants, i.e. algae, preceded animals and man.

Just as Moses closed the pre-Big-bang era with the appearance of light, demonstrating that the warming of primitive material had reached its peak before it exploded, here he used the coming of rain before the appearance of life to demonstrate that the earth's environment could now permit the blossoming of life from the level of chemistry that water had reached.

This was confirmed in this quote from the scientific documentary "Cosmos

– Beyond the Big Bang" written by Matt Hickey : "Underwater, mysterious chemical reactions come to create life". A sentence that I like very much, spoken by a scientist and demonstrating the limitation of science when approaching the physics of creation; a branch of astrotheology that includes God, the infinitely small, Rev. Dr. Sun Myung Moon's Principle of Creation and quantum mechanics.

4.2. Aquatic beings

In the scientific documentary "The Origins of the Earth" written by Alice Harper and Nigel Henbest, we are told that the appearance of fish was followed by flying insects. The Bible, speaking only of purpose, mentions winged beings instead, demonstrating that science has reached it. Let's read :

Genesis 1 : "*20 God said, 'Let the waters swarm with*

living creatures, and let birds fly over the earth toward the expanse of heaven. 21 And God created large sea- creatures and every living creature that moves, and swarms in the water, according to their kinds, and every winged bird after its kind. God saw that it was good. 22 God blessed them, saying, "This is good : Be fruitful, and multiply, and fill the waters of the seas, and let the birds multiply on the earth. 23 And there was evening and them morning : the fifth day".

In principle, all sea grasses are seaweed. These algae (cyanophyceae), at the appearance of life, have mutated, some into plants, phytoplankton, and others into marine beings, zooplankton, as we have said.

We know that the animal kingdom draws its food from plants and either inspires the gas they emit, or both at the same time, like man. Thus, the biblical logic of creation is an absolute and unquestionable truth. But by kneading the two verses above, we conclude that plants and water animals derived from algae all appeared on the fifth day of creation, according to the Bible, and in the fourth era, according to science.

4.3. Flying beings

According to the Bible, another significant point is the appearance of fish, followed by the appearance of birds. Indeed, this is confirmed by the existence of water or aquatic birds, including the emperor penguin, which dives more than 500 m in search of food.

But let's see how a biochemist explains this transition :

"The first multi-cellular life forms were algae and sponges, followed by jellyfish, worms and starfish. The first animals with a backbone were fish. Some of them evolved into amphibians. These gave birth to reptiles (crocodiles, snakes and lizards). The largest among them was the dinosaur.

Some small dinosaurs adorned themselves with feathers and evolved into birds, while another group of dinosaurs gave birth to mammals, which to this day remain the most evolved animals on the planet".

(Cf : Linda Gamlin, The Origins of Life, ed. Artis-Historia, Brussels. 1989, P.3).

Here we have the scientific scheme that led from underwater beings to birds, whereas the Bible only speaks of very important stages that are understandable to everyone. Speaking of underwater beings, it includes the evolutionary stage from the first forms of underwater life to birds, the last species of flying beings to appear. For example, how could Moses talk about certain beings that people had no knowledge of? So, for the Universal Cosmological Model, every finite and non-transitory form was the culmination of a principal.

Section 5 : from mammals to humans (sixth day)

5.1. The cohesion of the biblical pattern of creation

In our leading scientific documentary for this point, "The Origins of the Earth" written by Alice Harper and Nigel Henbest, we are told that between fish and reptiles, there were first flying beings. This also fits in with the biblical verse where reptiles are not found on the 5th biblical day like fish and flying creatures, but on the 6th day (Gen 1.25).

But which reptiles are we talking about here, since in the quotation below, from the biochemist and biologist Linda Gamlin, reptiles that evolved into dinosaurs are said to be the ancestors of birds ? Does the Bible contradict science ? Not at all, because in the last sentence of the same quotation we find another category of reptiles (dinosaurs) as we can read :

"*Some small dinosaurs adorned themselves with feathers and evolved into birds, while another group of dinosaurs gave birth to mammals, which to this day remain the most evolved animals on the Planet*".

(Cf : Linda Gamlin, The Origins of Life, ed. Artis-Historia, Brussels. 1989, P.3).

The Bible, taken literally, only contradicts science on evolution by expressing the breaking of symmetries with the words "God said, God did and God created". But the pattern of the appearance of things or beings in broad outline is almost in harmony between science and these writings of 1500 BC. The purpose of Moses, in writing the creation story, was to make the Israelites see that God is the master and Creator of all things, in as much as it is he who created the original atom "waters" as scientifically attested :

"Today, there are at least 4 million known species of living organisms on earth. They all use the same genetic code and their basic chemical composition is identical, which seems to indicate that they all descend from the same primitive species of bacteria, which were alive more than 3 billion years ago".
(Cf : Linda Gamin, Les origines de la vie, ed. Artis-Historia, Brussels. 1989, P.24).

Since this book reassures researchers about the truthfulness of the biblical accounts of creation, it is up to them to start revising their calculations in order to reach the divine order.

It should also be pointed out that the Standard Scientific Model does not detail all these points. Hence the Universal Cosmological Model, which is richer in opening up research from before the Big Bang to man.

Let's go back to our material and see what happened on the 6th biblical day or the 5th era according to science :

Ge 1 : "*24 God said, "Let the earth produce living creatures according to their kinds : livestock, creatures that crawl, and the wildlife of the earth according to their kinds... And it was so. 25 And God made the wildlife of the earth after their kind, and the livestock after their kind, and all the creatures that crawl on the ground according to their kinds. And God saw that it was good.26 And God said, Let us make man in our image, after our likeness, and let him have dominion over the fish of the sea, and over the birds of the sky, and over the livestock, and over all the earth, and over every creeping thing that crawl on the ground. 27 God created man in his own image, in the image of God he created him ; male and female he created them...31 And God saw all that he had made, and behold, it was very good. So there was evening and there was morning : the sixth day".*

In the sixth period of creation, called the sixth biblical day, we see here the creation of the beings which are very close to us, namely the earthly animals.

If the Bible had been written at the time when man made his discoveries, it would have looked like it was the result of a human calculation, based on the dating of fossils. Moreover, today we have a specialization of intelligence, so that the biblical creation story should call upon several scientific disciplines for its cohesion.

Even with a specialization of intelligence, today the mystery of creation still remains. Now, in the story of Genesis, the same person "Moses" was able to speak about everything : from the Pre-Big-bang to the appearance of man, in a coherent way. If someone still finds an inconsistency between the biblical scheme of creation and science, it is because he also has to reread the previous volumes.

5.2. The structure of living beings

In Ge 2:7 we read that man and animals were formed out of the dust of the earth (Universe), which science could never deny, because by analyzing the dust of comets, the molecules of life are constantly being discovered. But these molecules cannot interact with each other without liquid water with chemical proportions such as the water that is on planet earth.

Thus, before the creation of humans and animals, Moses began by first mentioning the problem of water.

In Gen 2:5" no shrub of the field had yet grown on the land, [d] and no plant of the field had yet sprouted, for the Lord God had not made it rain on the land, and there was no man to work the ground. 6 But mist would come up from the earth and water all the ground. "

Gen 2:7 "*The LORD God formed man from the dust of the earth*".

Even the Prophet Muhammad (peace and salvation be upon him) said that man was formed from mud, meaning the contribution of water in the dust. Once again, we discover a unique source of inspiration between science and religion, namely God.

5.2.1. Earth dust or star dust

Ge 2:7 "*And the LORD God formed dust from the ground, and breathed into his nostrils the breath of life, and man became a living being.* "

According to astrophysicists, man is formed from star dust. Where do they get this conclusion from?

Living beings are formed from the molecules of life. These have nebulae as their primary source. In the nebula, we find gas and dust. It is this dust that will one-day form the molecules of life in the stars, according to science.

Let's go back again to the processes of material creation. Indeed, like all material, the process of the creation of man, for scientists, began with the fusion of particles until they formed hydrogen at the beginning of the Universe. Hydrogen, upon heating, eventually decomposed into helium. The repetition of this type of process - allows the creation of the x's of the different chemical elements known up to iron.

All these elements, thus created, are found in the form of fluids and plasmas and are held by gravitation under a ball that we call the star. Iron is generally considered to be the final stage in the chain of nuclear fusion reactions within stars.

By exploding, through a phenomenon called supernova, the outer layers propel into space elements they have made inside them and gases we find in nature, including molecules of life.

Eventually, some of the material ended up forming the planet Earth. And it is in the planet Earth that man will eventually be formed by the elements that make it up, which Moses called the dust of the earth.

Scientists referring to the origin, i.e. the star ball, say that man is made of star dust. The Prophet Muhammad (peace and salvation be upon him), knowing that the peculiarity of life on planet Earth was due to the presence of liquid water, will say that man is made of mud; that is to say, the dust of the earth plus water. And, the last great prophet to think about this material, namely the Rev. Dr. Sun Myung Moon, concluded, in the Divine Principle, that man is the microcosm of the cosmos.

Thus, Moses resorted to the source to speak of dust as an element that has served to create life, starting with plants. Once again we see that science and the Bible speak the same language. But, to confuse this discovery by bypassing the Bible, scientists have preferred to speak of man as being formed from stardust.

Knowing that it is God who led all the processes from the origin to the formation of man as the last stage of creation, man finds himself to be the son of God and the Lord of the Universe. (Gen 1:27).

5.2.2 The breath of life in living beings

In the process of evolution, the difference between other living beings and man came from the divine aspect, or the consciousness attributed to each being, defined as the breath of life, an unfathomable field for science. Let us read :

Ge 2:7 *The LORD God formed man out of the dust of the earth, he breathed into his nostrils a breath of life and man became a living being.*

This verse cannot be interpreted as if God had blown into our nostrils the air we breathe.

Until then, the believers, in their literal analysis, thought that the breath of life was equal to the air. However, air is physical; it is the name given to the earth's atmosphere, which is composed of several gases. The breath of life, here, is part of something spiritual that cannot be felt. It is in fact the energy that is endowed to each being, as we have said, starting with the original particles that connect us with God. It is also the source of consciousness.

What God has blown into man has nothing to do with physical air. This breath has remained until today an enigma for those who wondered about things that cannot be felt in an individual, mainly the spirit and the soul. It is also through this that consciousness entered into inert material.

Moses describes it as the breath of life which is energy from God that the Moon couple explains in the book of the Principle of Creation and which is called the Universal Prime Energy and which was endowed to every living being and man was used as a case in point.

By saying that man was inert, and that the breath of God gave him life, we find ourselves in a phenomenon that scholars can describe as "the entry of energy into the human being". In the Divine Principle of the Moon couple, who are the first to explain the power of God in a scientific way, this sentence is articulated as follows :

"The First Universal Force has its origin in God and constitutes the force that God projects in each being at the moment of his creation" (Cf. : The Divine Principle, level 4, P.21).

As I keep saying in my textbooks, biblical language is spiritual; the physical term is used to teach believers. But the real explanation is to get into the heart of the one who said it.

Having exposed all this, we define the Universal Prime Energy, or the breath of life, as the energy of life. Without this energy, life becomes inert. This energy did not begin with the first molecule, but long before the Big-bang. It comes from God Himself. When we see a living being, we feel that it has energy, we feel that it has life, but we don't see this life sustaining energy. Its manifestation within the being creates the First Universal Force.

Section 6 : The mysteries of women's creation

Ge 2 : 7 And the LORD God formed man out of the dust of the earth, and breathed into his nostrils the breath of life, and man became a living being.

Ge 2 : 18 And the LORD God said, It is not good for man to be alone ; I will make him a helper like him.

Ge 2 : 21 Then the LORD God caused a deep sleep to fall on the man, and he fell asleep ; and he took one of his ribs, and closed the flesh in its place. 22 And the Lord God made a woman of the rib which he had taken from the man, and brought her to the man. 23 And the man said, this one, at last is bone of my bones and flesh of my flesh : This one will be called Woman, because she was taken from man.

Is there a scientific model that can start from the Pre-Big-bang and reach this level by talking about the mystery of women's creation ? And yet, the biblical scheme of creation closes up to this verse and if we do not touch it, the shell of our model, the Universal Cosmological Model, would be incomplete.

This is how what I have set myself as a task leads me to think beyond what religion and science can imagine before they take it over. Not being a biomedical specialist, here I bring only the snippets. It is up to the scientists to continue the appropriate research.

After creating man, the Bible tells us that he was alone and that God resolved to make him his complement. According to Moses, who is literally the author of this story, this complement was taken from the organs of the same man. Would an omniscient God forget, after having created all the male and female animals, to do the same for man and then remember it ? Scientifically speaking, is this process feasible or is this story hiding another secret that no one has been able to discover since it was written ? We will analyses the first possibility and answer the second, which will give an answer to the first question.

We are addressing this point because it is part of the biblical creation story and no one so far has dared to talk about it. Religious people sometimes use it to demonstrate their pre-eminence in front of women, but for scientists it is just an unfeasible myth that cannot even be looked at. Let's read :

Ge 2 : "*18The LORD God said, It is not good for man to be alone ; I will make him a helper like him. 21 Then the LORD God caused a deep sleep to fall on the man, and he fell asleep ; and he took one of his ribs, and closed the flesh in its place. 22 And the Lord God formed a woman from the rib which he had taken from the man, and brought her to the man. 23 And the man said, This is she who is bone of my bones and flesh of my flesh. This one will called Woman, because she was taken from the man. 24 This is why a man leaves his father and mother and bonds with his wife, and they become one flesh*".

Other Supporting Verses :

1Cor 11:8 *For the man was not taken from the woman, but the woman was taken from the man ;9 and the man was not created for the woman, but the woman was created for the man.*

6.1. Cloning

The story of the appearance of woman in the biblical story of creation resembles what science calls cloning.

Cloning is defined, in Genetic Engineering, as "*a genetic manipulation that makes it possible to obtain, from an original organism, one or more organisms with the same genetic heritage as the original organism*" (1).

(1) "cloning". Microsoft® Encarta® 2009 [DVD].

Microsoft Corporation, 2008.

In other words, cloning is the name that science uses when one living being is derived from another through genetic manipulation as defined above from Encarta.

Scientifically speaking, we can say that the history of the creation of women was the first genetic manipulation. According to biblical writings, everything happened as it does in our hospitals, even if it was not so. On the other hand, no religious can condemn me for commenting on it.

The history of cloning goes back to 1952, while the writings of Moses go back several centuries before Christ. Therefore, the man of today, whatever his intelligence, should keep his humility by recognizing the prowess of his creator and this great man of God who is Moses. Today, cloning has a pejorative meaning and is condemned by all religions as well as by the UN.

The practice of cloning is condemned exclusively by religion because it is said to be the exclusive preserve of the creator who is God. Because of their imperfection, men risk creating something that will be subject to perpetual suffering, or disability. Thus, on 8 March 2005, the United Nations General

Assembly decided to prohibit it, even for therapeutic purposes. But since this was not passed as law by the States, the number of cloned species, either for therapeutic purposes, or for commercial purposes, such as horses, or for the pleasure of passing from a wild breed to a domestic breed, or simply out of curiosity to see what the result is, is only increasing.

The biblical specificity about cloning is that it has produced an opposite sex; something that science has not yet achieved.

6.2. Cloning practice

The practice of cloning a man to have a woman, in the Bible, is not far from that of science, I had said. From what part of the man's body was the organ removed to create the woman according to the definition of cloning? Let us read the answer :

Ge 2:21" *Then the LORD God caused a deep sleep to fall upon the man, and he fell asleep, and took one of his ribs and closed the flesh in its place.*"

Our study concerns creation and content. As far as creation is concerned, when we say man or woman, we first see the sexual organs. But South Africa has gone too far, to the point of studying the testosterone of its Olympic champion Mokgadi Caster Semenya (born on 7 January 1991 in Pietersburg).

The body and face may be the same in both sexes, but it is the sexual organ that determines each person's identity. Even though transgender people are against this view, my thoughts on who an individual's identity belongs to was joined by Mr. Donald Trump, President of the United States, elected in 2016, a champion of extreme decisions.

Indeed, his administration wanted to put some order in this area, as we can

read :

"*According to the Trump administration, it is time for a definition of gender with a clear biological basis, based on science. The department would primarily want to rely on the genitals of the individual at birth, and genetic testing would be used in cases of uncertainty*". (Cf. : trustmyscience.com/author/jpierre/par le 24 octobre 2018).

To be clearer, according to the Trump administration, membership of a sex cannot be decided by the individual himself, whatever their body shapes. One should base oneself on the active sexual organ and not on the shape of the breasts or face.

Going back to the Bible, is it the forming cells of the sexual organ in women that have been amputated in men? It is up to science to answer this question, because when this organ is amputated, we lose part of our identity.

Moses, in the biblical story of creation, restored in the Universal Cosmological Model, tells us that after having anaesthetized man, God operated him and took one of his ribs (Gen 2 : 21-22). The term one of his ribs is to tell us that it was an internal organ protected by the ribs. And moreover, it was to draw our attention to the place where the first separation must have taken place. Which organ is it? Is it related to the reproductive organs? I have not yet finished my research in astronomy to go into medicine. But as no one has tried to think about these verses before me, I still draw attention to them so that I am not told that I have left out certain points of the biblical creation story.

Let's look at these images first :

Reproductive apparatus of the woman seen from the front

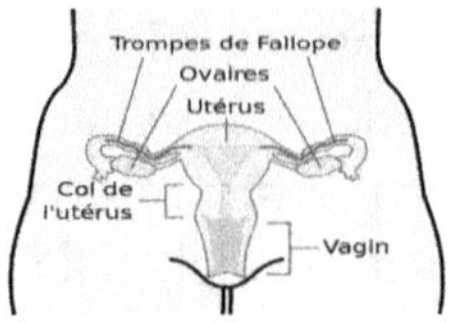

Reproductive system of the woman seen from the side

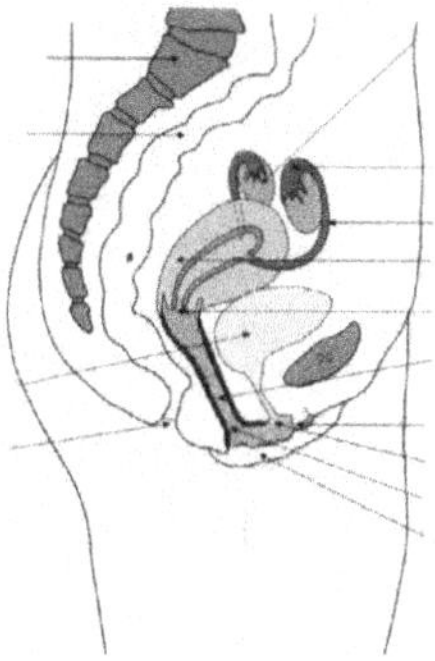

Human Reproductive System

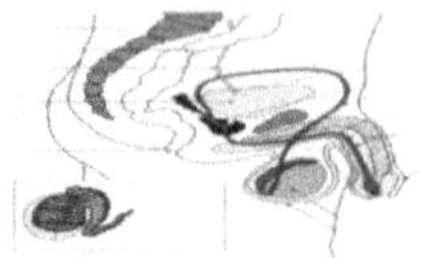

If the biblical story of creation were taken into consideration by science, the story of the creation of woman from man would already have been the subject of research. And science would already have come to the result that the two apparatuses are combined from the convex and concave parts. One receiving the other and vice versa to become one person.

If I start to bring the two patterns together, I will end up with a discovery that medicine has not yet made, but I prefer to specialize in cosmology. It's up to biologists to do their job. The understanding of the scientific details of the Biblical Model is now becoming the objective of all sciences. All I can say about this pattern is that what is concave for women, it is convex for men and vice versa. Thus, in order to be complete, these two parts always seek to complement each other.

For example, above the yellow image of the man, we see that the uterus with all its components is missing. Some of the functions of these components are performed by the man's testicles, placed at the bottom.

In fact, what I know is that it is in the reproductive system that we find the hormone called oestrogen which ensures the development of female characters in women, and in men, testosterone defines the male characters. In order to know these things several centuries before Christ, one had to be a great doctor or a scientist like Moses, who had direct contact with God, to understand the true technique of Genetic Engineering.

There are several forms of cloning. Let us see how cloning is done between two beings or within the same being to understand that the term "rib" was appropriate. The ovum is taken from the ovary of the female by opening the rib, as Moses said when medicine was not yet sophisticated. The nucleus is removed from the cell inside and the removed nucleus is replanted into the cell of the other animal or the animal itself. Once fertilization has been received, the egg is put back in its place.

Therefore, an animal generated by transferring nuclei from an adult cell into an egg cell emptied of its nucleus is called a clone. In this way, the same animal is resuscitated, but modified by the contribution of elements from the surrogate mother.

Through this experience, one would be tempted to say that Adam was not just man when he was created alone, because the new being can only grow in the womb before birth. But I am not going to discuss this in my book.

According to Rev. S.M. Moon's speech "True family and true universe centered on true love", 99,999 per cent of the baby's existence comes from the bones, blood and flesh of its mother, combined with the 0.001 per cent of its father's sperm. This is why the African ancestors, although not familiar with modern science, incriminate the woman more in cases of infertility of the couple. The man sows the seed and the woman is the fertile soil that provides all the nutrients. Man's infertility being a relative phenomenon.

What is disturbing with the creation of Eve is that Adam, being male, had an X and a Y chromosome ; Eve had two X's. This shows that all the genetic information necessary for the creation of Eve was in Adam. So Eve could be formed from Adam, but Adam could not be formed from

Eve.

In fact, a woman who would reproduce by parthenogenesis, that is to say without the help of the sperm, would only have daughters. Therefore, the carrier of the second life was drawn by God into the being carrying two different chromosomes. And Moses limited himself by saying quite simply that even Adam, in full anesthesia, completely ignored what had happened. Let's read :

Ge 2 : 21" *Then the LORD God caused a deep sleep to fall upon the man, and he fell asleep; and he took one of his ribs, and closed the flesh in its place.* "

It is in everyone's nature to go in search of what they have lost, what makes them incomplete - the opposite sex. At the sight of Eve, Adam exclaimed :

"*This is she who is bone of my bones and flesh of my flesh*" (Gen 2:23).

If the female had been created independently of the male, her presence would be indifferent to the male. Thus, no material how powerful or wealthy an individual may be, he can only live happily and happily by being next to a partner of the opposite sex, which is the answer of his heart. Hence the biblical verse :

Pr 18:22 "*He who finds a wife finds happiness; it is a grace from the Lord*".

But King Solomon complained, saying that out of a thousand women, he had not found one who was a bearer of happiness, i.e. who could make a man happy. But out of a thousand men, he found at least one who could make a woman happy. Perhaps if Solomon had gone to the level of a thousand and one, he could have found the woman who deserved the attribute of woman. The Bible says that finding such a woman is a grace from the Lord (Prov 18:22). This is simply to say that it is very, very difficult to find one.

If this is the case, then why he is married? This was the question Jesus' disciples asked him. Or in other words, why is it divorce if the next one is not one of the thousand exceptions?

It is also a grace for those who read my books to discover verses they did not know existed. Let us read :

Ecclesiastes says, "*This is what I have found, looking at things one by one to see why; 28 this is what my soul is still searching for, which I have not found. I have found one man among a thousand, but I have not found one woman among them all.29 But this I have found, that God has made men upright, but they have looked for many deviations.*

Matthew 19:10 "*His disciples said to him, 'If this is the condition of a man towards a woman, it is not good to marry.*

The first verse is to specify that God has nothing to do with the perverse characters of men (and women), what they do with them of their own free will. Jesus, on the other hand, does not deny the conclusion of the apostles in the second verse either.

Another reason for the words "part of his ribs" (Gen 2:21) is that the rib is the only human bone that regenerates and can be used to make another organ, such as a natural skull. This is where we see Moses' intelligence.

6.3. The hidden side of the women creation's story

So far we have been talking about the creation of woman according to the Bible. But, on closer inspection, we see that this story hides an enigma that we want to remove.

When we ask ourselves why man was created alone, according to the Bible, when he was the last creature, we come to the conclusion that this happened long before he was formed and that it does not directly concern human beings. If God had only realized at the end of creation that the opposite sex to man was missing, then he would have had to start all over again from man to microscopic beings. Now, the creation of man brought God's work to a close as it is written.

Gen *1:26 And God said, Let us make man in our image, after our likeness...*

1:27 God created man in his own image, in the image of God he created him; male and female he created them.

1:28 And God blessed them, and said unto them, Be fruitful and multiply, and fill the earth, and subdue it.

1:31 And God saw all that he had made, and, behold, it was very good. And there was evening and there was morning : and it was the sixth day.

Ge 2:1 And the heavens and the earth were finished, and all the host thereof. 2:2 And God finished his work which he had made on the seventh day : and he rested on the seventh day from all his work which he had made.

As can be seen in chapter I, all the processes of creation were completed to perfection, ending with the words "very good".

While the details of chapter two, in which God is said to have found that man was missing something, are given in greater detail in chapter two.

Indeed, to explain a situation, Moses had to use, as an illustration or case study, visible things, taking into story the intellectual level of the people of his time. Through this diagram of Genesis, Moses wanted to teach us that the first living organism, without a predefined sex, had a male predominance before duplicating itself into two sexes. Moses could only use the names of man or Adam instead of male, and woman or Eve instead of female. Hence the phrase :

Ge 2 : 18" *The Lord God said : It is not good for man to be alone; I will make a helper like him*".

I had already said that chapter two of Genesis is a scientific illustration of creation. Had it not been for this, I would have had great difficulty convincing the scientists of the omniscience of God regarding the formation of living beings, since the formation of the Universe is clear.

Indeed, Moses would have been considered a madman to express the feeling of a bacterium or an animal that sees its opposite, with the phrase "this is bone of my bones, flesh of my flesh" (Ge 2 : 23). Let us take a male animal and keep it in solitude. Let us bring it to him afterwards, its female. The way he will leap after overcoming his selfishness will demonstrate this sentence of Moses, although he cannot express it verbally. Through this clarification, we can understand why Moses took "Man" as his case.

Just as in the beginning we spoke of a single continent, "Pangaea", which was divided into several continents, it is now up to science to join the Bible by studying in which cycle of evolution this separation of sexes occurred and why Moses spoke of the opening of a coast. This is an orientation for research. From now on, the Bible becomes a book which, with the publication of the books in the series "The Verses of the Scholars", begins to give researchers food for thought with a scientific scope. Let's go back to the beginning of evolution.

Indeed, like all living beings, bacteria are 95% composed of four natural elements : carbon, hydrogen, oxygen and nitrogen. (Cf. : J.B. Veyrieras, Science&vie Hors-Série n° 288, 09/2019).

In a way, they are only our microcosms and everything should be described from them in order to understand the hidden face of the story of the creation of the opposite sex from a single being.

Thus, to solve this enigma we have to resort to the

sexes of the first created beings, that is to say, bacteria and cyanobacteria. When I came to this conclusion, I didn't know the surprise that awaited me when I read the article below :

"Cyanobacteria appeared quite early in Earth's history, when conditions were not yet favorable for the life forms we know today. They do not reproduce sexually even though some genome exchanges are known. Their only means of reproduction is bipartition. Bipartition, in the general sense, which is little used today, expresses the separation of something into two parts, which then continue to exist independently."

(Cf. : futurascience.com/planets/definitions/botanical-cyanobacterium-122/")

So, the story of cloning women was to show us that the first living beings began without a predefined sex, multiplying by bipartition before duplicating in two by going through different stages, from asexual to sexual. This bipartition, according to my understanding of the biblical verse, was done by the coast.

Here is an example of this :

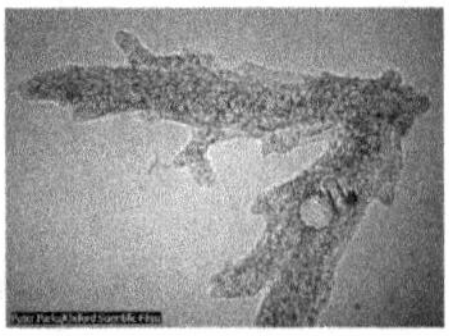

Amoeba in the process of division

Asexual reproduction by simple cell division allows single-cell organisms such as amoebas to multiply rapidly when conditions are favorable. In this mode of reproduction called scissiparity, the "mother" cell divides into two identical "daughter" cells. Peter Parks/Oxford Scientific Films

Microsoft ® Encarta ® 2009. 1993-2008 Microsoft Corporation.

Thismodeofself-reproductionofearlyorganismsiscalled Asexual Reproduction. It can be done in 3 modes that confirm the use of the word "rib".

– the individual literally splits in two to produce two identical individuals of the same size;

– a budding during which an extension of the organism separates and produces a new individual;

– one or the multinucleate parent divides into several cells of similar size.

(Cf. : .http//simulum.bio.uottawa.ca/bio2525/Notes/les-Protozoa.htm")

Here are the main parts of a protozoon

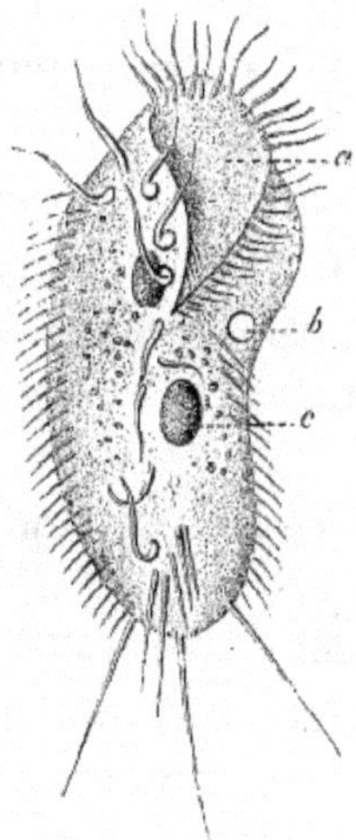

A ciliated infuser : Stylonichie moulds. a, peristoma. - b, vacuole. - c, nucleus,We read : *"The nucleus generally serves, as in any cell, to carry out mainly*

the functions of multiplication and reproduction". (Cf. :http ://www.cosmovision.com/protozoaires.htm").

When we see the location of the nucleus which can only be reached from the rib, we conclude that this name was indeed justified by Moses in relation to the other organs of the body. Let's read again :

Ge 2 : "21 *Then the LORD God caused a deep sleep to fall on the man, and he fell asleep; and he took one of his ribs, and closed the flesh in its place. 22 And the LORD God formed a woman from the rib which he had taken from the man, and brought her to the man.*

How did things happen from two identities to gender? Moses says that God had to take a rib out of man for an evolution towards two sexes in living beings. It is not only for the man that the woman is flesh of the flesh. Even for animals of the same species, the female is also flesh of the male. Therefore, things must have started before the creation of man.

Man was the summation of all that was created before him, that is to say, a miniature cosmos that we call a microcosm. It was considered by Moses as a case study to help his contemporaries understand what the beginning of the history of living beings was, because, as I said in the text, it is only man's feelings and his voice that can be expressed in writing.

So one of the second greatest discoveries for biologists is that the first being did not have a well-defined sexual classification. So we went down to the level of cyanobacteria and found that they multiplied by bipartition, confirming the Bible.

Even speaking of the deep sleep to which God subjected man to operate, this is not to be confused with our natural sleep. It is rather the time it took to pass from the asexual to the two sexes, each with its own identity.

For those who are still thinking about human cloning, science has recognized that it is materially very difficult to conceive and this is because of the large number of mature eggs that it requires, whereas women have a limited number. We have used the word "difficult" because nothing is impossible against man's evil ingenuity, which God has established as the Lord of creation (Gen 1:28).

When we do not know how to explain why a pineapple planted in the same bed with a pepper (pimento) have different tastes, then we will have to accept the limitations of our thoughts without denying the search for general knowledge.

Chapter VI :
The fulfilment of the predictions of the Universal Cosmological Model

The biblical story of creation, called the Biblical Model in Cosmology, which is rendered in the Universal Cosmological Model in the updated language, dates from the 15th century BC. A cosmological model is only taken into story if what it predicts is proven in theories and observations. The dazzling development of astrophysics from the 20th century AD onwards makes it possible to confirm the Biblical Model as given in the Universal Cosmological Model.

Section 1 : Pre-Big-bang

1. A formless and empty earth

Ge 1 : 2 " The earth was formless and empty ".

A formless and empty earth is what the biblical scheme predicts for the beginning of the Universe. Scientifically, this would mean that the Universe began in a vacuum. I thought that I was the only one to defend this point of view in the Universal Cosmological Model, but what was my surprise when I discovered that it was also part of the String Model.

So, the integration of the Uncertainty Principle in the String Model makes it possible to explain the evolution of the pre-Big-bang Universe in a scientific way as literally developed in the Universal Cosmological Model.

2. Darkness and the Abyss

Ge 1 : 2 " *The earth was formless and empty : there was darkness on the surface of the abyss* ".

.2.1. Darkness

The Universe is dark as the biblical scheme states. In Olbers' paradox, Heinrich Wilhelm Olbers (1758-1840), a German physicist and astronomer, asked himself the question, why is the Universe dark? In 1826, he examined a problem already commented on by thinkers before : if the Universe is infinite and homogeneous, with stars in all directions, the night sky must be uniformly illuminated. This observation, known as Olbers' paradox, was solved, according to scientists, when it was discovered that the observable extent of the Universe is probably limited to a radius of 20 billion light years (Cf : "Olbers, Heinrich

Wilhelm Matthaus "Microsoft® Encarta® 2009 [DVD]. Microsoft Corporation, 2008).

Within the framework of the Universal Cosmological Model, we can add that the expansion of the Universe, according to science, is taking place at a speed greater than the speed of light. Thus, since the Universe is naturally dark, according to the biblical scheme of creation, its large size does not allow the stars that are born and die to bridge this gap, which is only increasing, nor to illuminate it.

Furthermore, we have said that the expression "biblical darkness" also refers to an energy that can only be dark energy.

2.2. **The abyss**

The Universe came out of chaos, this is the conclusion of the philosophers. But in the physical sciences, we speak rather of the vacuum and the collapse creating either the black hole for some, or the deformation of the fabric of space-time by the condensation of particles, according to the String Model. This is the common structure for all those who think about the Universe Pre-Big-bang. This is also predicted in the biblical scheme through the verse "there was darkness covered the surface of the watery depths " (Gen 1:2).

2.3. **A wind of God or a spirit of God**

Gen 1:2 "...the Spirit of God moved over the waters...".

To acknowledge that there was a pre-Big Bang is to acknowledge that we were dealing with a dynamic Universe. This dynamism was taking place at the particle level, as the String Model and the Biblical scheme attest. This is reported in the Bible by the phrase "A wind of God was moving over the waters" (Bible Jerusalem Version).

According to the Universal Cosmological Model, this dynamism was characterized by cosmic deflation until the condensed Universe was created. The uniform temperature of the cosmological diffuse background in all directions of the sky confirms that at a certain point in its evolution, all directions benefited from the same heat. This is possible in the case of a condensed Universe described as water by the Bible and the original atom by the Belgian scientist Georges Lemaitre.

2.4. **And there was light**

Ge 1:3 "...God said, let there be light. And there was light.

Confinement leads to the heating of material. In the pre-Big-bang period, the situation was characterized by a deflation of the Universe causing particle- anti-particle collisions. It's up to astrophysicists to study and tell us what we've come to, until we get to the condensed Universe.

During the condensation cycle, the light that the Bible speaks of "*God said, 'Let there be light, and there was light*', was produced with extreme heat.

This can be understood by this law of physics : "According to the laws of thermodynamics, this contraction (of material somewhere in the cosmos) is necessarily accompanied by a significant rise in temperature at its centre and a strong release of light energy (Cf. : Ciel&Espace no.370, March 2001. Les feux du ciel, P.64, by Leila Haddad).

The scientist Roger Penrose, in his model of cyclical universes, explained the emergence of a light before the Big-bang, described as an "incident light ray", while ignoring the biblical scheme of creation. We present this scheme here, supplemented by our comments indicated by the arrows.

Minkowski's diagram of space-time

(Diagram of Penrose : Chaours.rv.pagesperso.orange.fr/physics/Stars/Black).

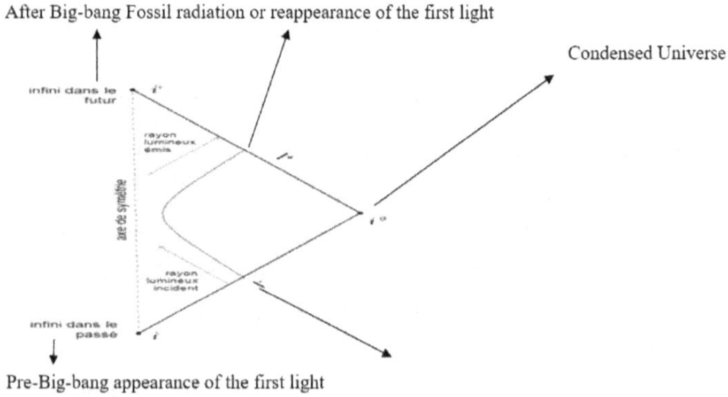

Ge 1:3 " *God said that there was light and there was light* " (Pre-Big-bang or first light).

At present, we can say that it is this light that is part, to this day, of the famous fossil radiation.

Being confined to the extreme, at point Io, even the light could not escape. It was only after the Big Bang, 380,000 years later, that there was light able to reappear under the name "emitted light ray".

All the hypotheses, evoked until then, of a light that originated through recombination, cannot stand upright because if this were the case, this light would have to activate with the many supernovae known to the Universe and would no longer have to remain, to this day, in a statically uniform value in every corner of the Universe. If it is through the recombination of atoms after the Big-bang, it should already disappear. It is by thinking of an explosion of the condensed Universe that Georges Gamow must have predicted it.

Moses had described this light as "Daylight" to demonstrate its eternal character, thus preventing the Universe from sinking into absolute zero by a physical mechanism not yet detected in theory.

Several astrophysicists have already put forward the hypothesis that the instability in the image of the cosmic background, with spots that cannot be explained, may have originated before the Big Bang.

2.5. **And the light separated from the darkness**

Gen 1:4 "*And God saw that there was light; and God separated the light from the darkness.*

For the common man, this sentence does not even merit meditation, but it is not the same for an astrophysicist. In everyday life, as soon as the light appears, the darkness disappears. But here, in the emergence of light, there was an interaction with darkness, demonstrating that we were faced with two energies which acted differently or that there was a third element, also invisible, which separated the two. This is how we also said that dark material finds its source in the Pre-Big-bang.

We were surprised to find that an astrophysicist, namely Mr. Tommi Tenkanen, from the Johns-Hopkins University, made the same hypothesis, in scientific terms, that we can read :

"*If dark material was really a remnant of the Big Bang; in many cases, researchers would have already seen a direct signal from dark material in different particle physics experiments*". (Cf. : trustmyscience.com/new-ear-dark-material-more-ancient-than-big-bang/.by Stéphanie Schmidt, 9 August 2019).

Therefore, he concluded quite simply that dark material should be from the Pre-Big-bang.

For us, this scientific article also confirms the Universal Cosmological Model and now calls on researchers to look into it.

So our support for the existence of dark energy and dark material comes from the Biblical Model and not from the arguments of scientists. Even saying that the Universe began in the vacuum, we first discovered it in the Bible and supported it, before I discovered it in the String Model.

As I said, this verse is meditative for astrophysicists and is a line of research that can teach us more about dark material and dark energy.

Section 2 : The Big bang

Ge 1:6" *God said, "Let there be a expanse between the waters, and let it divide*
the waters from the waters".

This verse makes us see that we were dealing with something unique called "waters" that should be separated.

It was around the 1920s to 1930s that the idea of a Universe that began as an atom and then expansion was born. The authors of this theory were the Russian physicist Alexander Friedmann and at the same time the abbot astronomer Georges Lemaitre. Their prediction was brilliantly confirmed in 1929 by the observations of distant galaxies made by the American astronomer Edwin Hubble.

But it was not until the theory developed by the Russian-born American physicist George Gamow in the 1940s that a physical description of the first moments of the Universe was obtained, predicting an extremely dense and hot primordial phase which led Hoyle to ironically call it the "big bang" (see "Big bang, theory of the" Microsoft® Encarta® 2009 [DVD]. Microsoft Corporation, 2008).

The expansion of the Universe, the 3 K radiation of the cosmological diffuse background and the possibility of explaining the presence of light elements have often been presented as the three proofs of the big bang.

Moses used the term "separation of the waters" to refer to the Big-bang. When in the Pre-Big-bang, the water was contracting by a gravitational force causing the deflation of the universe, he said "a wind of God was moving over the waters" (Gen 1:3). After its extreme contraction and before recognizing the rebound, he said, "God said, Let there be an expanse between the waters, and let it separate the waters from the waters" (Gen 1:6).

And after he has Big-bang, he will say, ”And God made the expanse, and separated the waters that were under the expanse from the waters that were above the expanse. And it was so” (Gen 1:7).

And after the Big Bang, when the material was created, he will conclude by saying, ”God said, ’Let the waters under the heaven be gathered together in one place, and let the dry land appear. And it was so. (Gen 1:9). God says : Let there be lights in the expanse of the heaven to divide the day from the night ; let them be for signs to mark times and days and years (Gen 1:14)”.

Therefore, everything was formed from the material that he originally described as water. This demonstrates the greatest proof of the Big-bang as scientists have found through this quotation : ”Moreover, observations show that on a large scale, material is distributed rather homogeneously. (Cf. : Ciel&Espace n°370, March 2001, serge Jodra, P .43).

The observations made on the composition of material in the Universe and the discovery of fossil radiation have confirmed this Big-bang theory, described in the Bible as the ”separation of the waters from the waters”.

Section 3 : After the Big-bang

3.1. The formation of the solar system

Ge 1 : 9 ”*God said, ”Let the water under the sky be gathered into one place, and let the dry land appear. And it was so. 10 And God called the dry land earth, and the heap of waters he called seas. And God saw that it was good. 13 And there was evening and there was morning; third day*”

Ge 1:14” *And God said, Let there be lights in the expanse of heaven to divide the day from the night; and let them be for signs to mark times and days and years; 15 and let them be for lights in the expanse of heaven to give light upon the earth. And so it was. 16 And God made the two great lights, the greater light to give light by day and the lesser light to give light by night, and the stars. 19 And there was evening and there was morning; fourth day*”.

The Solar System is a planetary system formed by the Sun and the celestial bodies that gravitate around it, such as the eight planets, including the Earth. Before their age and the materials which they are composed were discovered in cosmology, these biblical verses were like a prediction, and I am the one who says so. Science, without taking them into consideration, discovered that all the planets in the system have the same composition. The same materials, which are in fusion and in the form of plasma in the Sun, are found on the Earth and on other planets in solid, liquid and gaseous form.

Moses, who in his time, 1500 BC, did not have the apparatus to understand the composition of the Sun and the stars, predicted something very great by saying that it was the same material, which he called water, that separated from the one below it was used to create the Earth and the one above it was used to create the luminaries. If for believers this is part of faith, to this day no astrophysicist can question this prediction, called prophecy in religion.

The other greatest prediction still concerning the solar system is the age of the stars within it. All the material that formed the stars of the solar system was, according to Moses' predictions, projected at the same time, but given the evolution of each of the stars in their orbits, it is the Earth, as he said, that formed before, in the third stage and the others in the fourth.

To date, it is attested that the solar system dates from around 5 billion years ago and that the Earth would be 4.6 billion years old, slightly older than the Sun, which is 4.5 billion years old. However, there are sources that give the Sun a slightly older age, but this does not material, especially since the solar system places them all at 5 billion years old. All this is evidence of the predictions of the biblical model given in the Universal Cosmological Model.

3.2. The appearance of living beings

3.2.1. From algae to birds

Ge 1:*20 And God said, Let the waters swarm with living creatures, and let birds fly over the earth toward the expanse of heaven.23 And there was evening and there was morning; fifth day.* "

Life has germinated in water and no scientist can doubt it any longer. What will astonish everyone, and which I developed in Volume II, is the fact that Moses closed this first era of the appearance of living beings by birds before the sudden end of the dinosaurs, which marked the beginning of the development of mammals. Can we classify dinosaurs in this era? Let's read the description by biochemist and biologist Linda Gamlin :

"*The appearance of life on earth took place under very different conditions from those we know today. The air contained little oxygen, erupting volcanoes and violent thunderstorms were commonplace. Lightning caused simple chemical substances to combine in the atmosphere, which then formed more complex chemical compounds - the basic elements of life*" (Cf. Linda Gamlin, Les origines de la vie, ed. Artis-Historia, Brussels. 1989, P. 4).

In the scientific documentary "the origins of the earth", written by Alice Harper and Nigel Henbest, the first beings were the cyanobacteria, some which are known to have mutated into plant algae that will evolve into plants on earth and others into fish. After fish, the flying insects came, followed by reptiles and dinosaurs, primates and finally man.

This scientific classification goes hand in hand with what Moses wrote in the Bible (Ge 1:20) more than 1500 years before Christ, but at that time there was no fossil study. So it was also a scientific prediction.

According to science, dinosaurs and reptiles were the masters during this first stage of life. They really appeared between the first underwater beings and birds, as biochemist and biologist Linda Gamlin concluded :

"*The era of reptiles ended 65 million years ago, at the end of the Cretaceous period, when a large number of species disappeared as suddenly as mysteriously, including the dinosaurs...The disappearance of the dinosaurs marked the beginning of an era rich in possibilities for mammals. They multiplied spectacularly, and many forms appeared* "(Cf. : Linda Gamlin, les origines de la vie, ed. Artis -

Historia, 1989, Brussels, Pp.14-18).

3.2.2. From mammals to man

Ge 1 : 24 *produce living creatures according to their kinds : livestock, creatures that crawl, and the wildlife of the earth according to their kinds. And it was so.27 God created man in his own image, in the image of God he created him, male and female.31 God saw all that he had made, and behold, it was very good. And there was evening and there was morning : the sixth day.*

Moses ended the fifth day with the appearance of the birds, before we enter the sixth day, which is from mammals to man. But, how was this transition made from a physical point of view so that it could also be considered a prediction that was fulfilled with the discovery of the fossils ?

According to the biochemist and biologist Linda Gamlin, amphibians evolved into reptiles during the Carboniferous period, some 300 million years ago, and these reptiles became more diverse. Some mutated into dinosaurs. Other dinosaurs were the ancestors of birds and mammals. A fossil known as Archaeopteryx has shown how birds evolved from dinosaurs.

(Cf. : Linda Gamlin, les origines de la vie, ed. Artis -Historia, 1989, Brussels, Pp 14-18).

3.2.2.1. Man was formed out of the dust of the Earth

At the time when the Genesis writings were written, 1500 BC, who could know the composition or constituents of the human organism ? Yet Moses was the first to predict that man was formed out of the dust of the Earth. To convince us of this, we only have to read the composition of a multivitamin drug. Everything we mention is part of the composition of our organism and it always needs it for its growth.

Astrophysicists, having observed that these minerals form in the stars, will all conclude that man is the product of stardust. The astrophysicists resorted upstream, while Moses resorted downstream. Thus, the language is the same.

So, from the Big-bang to the living, nobody will be able to question us about the predictions made by the Universal Cosmological Model. All that Moses predicted and which is recorded in the Bible, proved to be correct with the appearance of scientific proof.

As for the pre-Big-bang to Big-bang period, it is now that scientists have just become effectively aware of its reality through the series of books "The Verses of the Scientists" and that theorists, NASA and CERN will now set about bringing all the scientific clarifications related to this period.

Chapter VII : Forum

I have selected several questions related to the subject of this book, which were asked to me on the Quora.com website. On several occasions, I answered only partially because I had not yet published my books and I did not want to reveal my thoughts without foundation.

The majority of the questions concerned what was there before the Big- bang. This effectively demonstrates that the established models only partially quenched people's thirst. Even the theories that came close to the pre-Big Bang sphere have not been able to meet people's expectations. Science is not behind the times, but it has not yet reached the sphere where its research should lead, but it did not even know it, limiting itself to Planck's Wall.

With the advent of the "Universal Cosmological Model", science has just discovered that the Big-bang is only a transitory phenomenon that takes place biblically on the second day, and that the first day was also very eventful. With the restitution of the biblical creation story in the Universal Cosmological Model, answering someone about what was before the Big Bang becomes easy and even makes one want to popularize it.

It is not for me to undertake this popularization, because I have the language of Moses, hence I can only follow the advice of the philosopher Gilles Dowek, who wrote : "*What cannot be spoken about must be written down*". (Cf. : Gilles Dowek, Ce qui on ne peut parler il faut l'écrire, ed. Le Pommier, Paris, March 2019).

It is the astrophysicists themselves who, from now on, will begin to visit the different Churches instead of confining themselves to the planetarium for the popularization of the Universal Cosmology Model. It has nothing to do with a question of faith, but it is the restitution of the first cosmological book in which the author attributed to God the physical phenomena that have taken place since the creation of the Universe, and, as I said, it is the only model that makes it very easy to understand the Universe and that we can even write books for high school students.

If I build a building, no material how many floors it has or how big it is, or whatever is inside, neurobiologists know that in my brain the part that this plan occupies is just a speck of dust compared to all the information that the brain stores. And the phrase has always been "this is the building I'm building", even though it took a lot of machinery. Why did it take a lot of machinery? Because the plan is part of the fruit of my brain and the means came from me.

It's the same thing when you say that God is the creator of the Universe. But this was not done by magic because the bricks, called particles, were very real, described in the Bible as "waters". And it is this last part that interests the scholars, each one of them drawing his conclusion and saying whether God is the creator of the first brick or not.

As for me, since the functioning of the Universe respects the principle of the dual characteristics of God, I have come to the conclusion that He who is the creator of the cosmos.

I do not impose my faith on anyone, but I have achieved my goal, namely that of arriving, in an updated language, at restoring to the narrative the biblical scheme of creation.

The Universal Cosmological Model also brings an advantage to the standard model that it swallows up, in the sense that no one will be able to question the Big Bang any more. This word used to frighten believers, not knowing that biblically it was described by the terms "separation of the waters from the waters". It is now becoming their physical proof that God created the universe.

In this forum, each question has its author, which is why we can sometimes see the repetition of an answer, from one question to another.

So, let's hope that the answers given in this forum will bring a plus on the understanding of the Universal Cosmological Model.

Section 1 : From the Pre-Big-bang to the Big-bang

1.1 Is there something beyond the cosmos? Do the latter represent infinity?

The word 'cosmos' has two meanings :

– In science, the cosmos represents the observable Universe and the beyond, of which we do not yet know whether it is finite or infinite. The observable Universe is so far delimited by the cosmological horizon, which is about 45 billion light years away from us. This means that our current devices have enabled us to see the light that comes from this distance and after that there is only deep darkness, or darkness as the Bible says. It is this darkness, the extent which we do not know with the observable Universe, that we call the cosmos in science.

– For religion, the cosmos has another meaning. The Bible says that "in the beginning God created the heavens and the earth" (Gen 1:1). According to the scientific reconstruction of the biblical creation story in the Universal Cosmological Model, the heavens represent the spiritual world and the earth the universe. The two spheres together form the cosmos. Therefore, for the Bible, there is no beyond the cosmos.

1.2. Was the Universe created in six days as mentioned in the Bible?

Understanding the notion of biblical times is part of the basic foundations for understanding the Universal Cosmological Model. The Universe was not created in six ordinary days as read in the Bible. The six biblical days represent six major stages which the Universe created, from the beginning to man. This is not an interpretation since during the first three days of creation there were no reference stars to determine time. The reference stars were not created until the fourth day; the Bible itself confirms this.

In addition, there is a biblical verse that says that for "God, one day is like a thousand years" (2 Peter 3:8). It does not say that one day is equal to a thousand years. Therefore, a period can take as many years and be called a day biblically. To achieve only one goal, its achievement, from beginning to end, is biblically called a day. In this way we understand that these days are not equal, therefore they have nothing to do with our 24-hour days.

For example, only one event was described on the second day; that of the separation of the waters from the waters, or scientific Big-bang. Even science knows that this time lasted only a thousandth of a second. But since it was a momentous event in the history of the Universe, even Moses in the Bible gave it a day apart from other days when many events took place.

Therefore, as soon as a created thing came to maturity according to evolution, Moses interspersed it with a day. Thus, the six days represent the six great stages and periods of the creation of the Universe, up to man.

1.3 Is there a one in seven chance that the Universe began on a Monday?

Your question surely refers to the Bible, which speaks of six days for creation and the seventh day for God's rest. The Bible did not mention the word Monday. Religion, by taking up this notion of time, makes all Christians believe that it was the seven days of the week. But there is an Apostle who warned against this way of thinking by saying that "do you not know that for God, one day is like a thousand years and a thousand years like yesterday's day ?".

In the books of the series "The Verses of the Scholars", the notion of biblical times is widely developed. To believe that the Universe was created in seven days is a material of great ignorance. Let us quote a few sentences from Volume I : "To speak of fixed times, you need two stars, one which revolves around the other. However, during the first three days of biblical creation, there was no creation of stars. So what were these days fixed according to ?

It is understandable that they were not our twenty-four hour days, but they were defined periods of days to mark the accomplishment of one stage to the next.

1.4. How long did the Big-bang last ?

According to science, it lasted between 10^{-46} seconds (10 exponents -46s) and 10^{-43} seconds (10 exponents -43s) which marked the beginning of cosmic inflation.

But it was a very important moment, both for science and for God, so much so that Moses, in the Bible, assigned a whole day for this event. It was the second day on which he spoke of "the separation of the waters from the waters".

Indeed, according to the book "Cosmology according to Biblical and Ancient Models" published in July 2020, while scientifically restoring the biblical creation story, I was the first to discover that the Big-bang is indeed biblical and that the pre-Big-bang was a substantial reality.

1.5. Does anyone have a theory about what was just before the Big-bang ?

Yes, someone does have a theory about what was just before the Big-bang and it is found in the Universal

Cosmological Model developed in the books of the ”Verses of the Scholars” series.

1.6. How can we describe what was there before the Big Bang? Is the Big- bang the beginning or the end of a process?

The Big-bang does indeed mark the end of a process, i.e. the process before the Big-bang. Thus, it is still part of a field that is still impenetrable to science. The description of what happened before the Big Bang is recorded in the Bible on the first day of creation, and it is rendered in the Cosmological Model in actualised language.

To give a brief summary, the Universe, which was initially characterized by a vacuum, experienced a deceleration on the original particles until it condensed. As for the phenomena that took place between the beginnings of the deceleration until the condensation, this is described in the Bible on the first day and restored in detail in Volume I.

It was after this extreme condensation that there was Big-bang, whose subsequent phenomena are described in the Standard Model or Big-bang Model, completed by the Universal Cosmological Model.

From this, we can conclude that the Big-bang marks the end of a process, which was Pre-Big-bang, and the beginning of a new process, which is after Big- bang, as shown in this table :

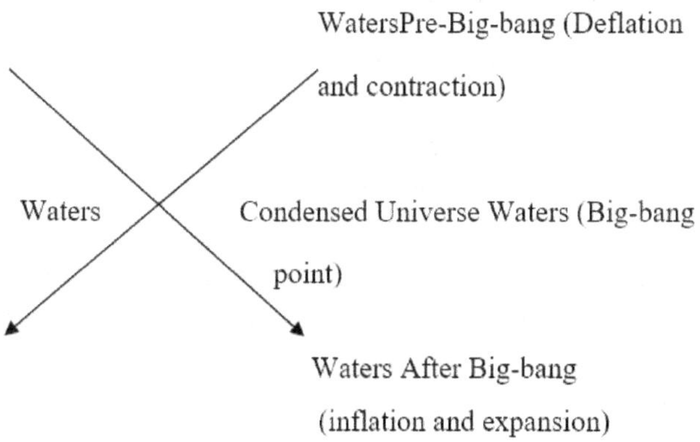

1.7. Does the Universe have a beginning?

The Universe does have a beginning. Several models tending to say that the Universe has not had a beginning have been swept away by science. The only question that remains is, when did the Universe begin? Planck's Wall provides a barrier that science has not been able to penetrate until now.

Is it necessary to resort to metaphysics to quench our thirst? What if we take a look at the first cosmological book par excellence, the Bible? That is indeed what I did. In my research, I made a study on "Cosmology according to the biblical and ancient model" which is also the title of Volume I and which I continue to develop in other volumes.

The result was that the Universe had a beginning which, after the process of cosmic deflation, resulted in a universe condensed to the extreme and it was after the Big-bang intervened.

1.8. What do we know about the Big-bang?

According to the "Universal Cosmological Model" which I am the inventor, and which is the only one to trace the situation from before the Big-bang to after the Big-bang to man, the Big-bang represents a stage where the Universe, after undergoing deflation in the pre-Big-bang, became extremely condensed before undergoing an expansion known as cosmic inflation. The transition from extreme condensation to expansion is what is known as the Big-bang.

1.9. Apart from the Big-bang what other theory exists to explain the creation of the Universe?

Until now, there has been no valid theory. Even the Big-bang theory does not explain the creation of the Universe. The only cosmological book that explains the creation of the Universe is the Bible. The Bible is written in a religious language that is very difficult to decipher. Everything is attributed to the divine creation without saying the physical processes that were involved. This is why a scientific restitution of this language is necessary if we want to advance in cosmological knowledge. This has just been done by me.

This restitution brings out the first cosmological model that really explains the creation of the Universe. It is the Universal Cosmological Model developed in my previous books in the series "The Verses of the Scientists".

However, in reading this volume, in the section concerning "The Universal Cosmological Model versus other pre-Big Bang theories", we have listed and explained the theories of the few authors and researchers who were opposed to the pre-Big Bang theories.

1.10. If the Big-bang created the Universe, what created the Big-bang and what created what create

the Big-bang?

The Big-bang did not create the Universe. It was a transitional stage in the Universe. According to the Universal Cosmological Model, which is the restitution of the first cosmological book par excellence, namely the Bible, the Universe being created religiously by God, and physically by virtual particles, their deflation, after several processes, caused it to be condensed. This is what caused the Big-bang that followed the moment of condensation.

So the Big-bang was not created, but was a consequence of the physical phenomena that science is trying to elucidate.

Also in the Bible, the Big-bang took place on the second day of creation and is described as "the separation of the waters from the waters". This gives us a direct understanding that the Big Bang did not create the Universe as some scientists would have us believe before.

1.11. If, in theory, before the Big Bang there was nothing, how could there have been an explosion? Something doesn't add up... Isn't the Big-bang finally the scientific reading of creation by a superior being who would have triggered everything?

Your concern : it's not what happened after the Big-bang, but rather how did we end up there. You are right to ask this question because we are told that the temperature at the time of the Big-bang was around 10 exposing 15°C. Nothing can begin an existence at that temperature. In conclusion, there had indeed been a Pre-Big-bang.

The study of what happened in the pre-Big-bang is contained in the books of the series "The Verses of the Scholars". To achieve this, we have reconstructed the biblical story of creation in the updated language, thus giving rise to the "Universal Cosmological Model". This old model, which has become new, is a scientific reconstruction of the biblical story of creation. This model, which can be described as the biblical model, has not yet been integrated by science.

Your second part of the question, namely : Isn't the Big-bang ultimately the scientific reading of creation? That's what this model answers.

1.12. If time did not exist before the Big-bang, how could the Big-bang take place?

Who said that time did not exist before the Big-bang? It was men like you whose mathematics could not solve the impossible. But if you ask the Bible, you will see that time existed. This question was done during the research that led to the development of the "Universal Cosmological Model"; which restores the biblical scheme of creation with a scientific language.

The Universal Cosmological Model, developed in the books of the "Verses of the Scholars" series, reveals to us what the spirit of God, or a wind of God, means in a scientific way and what these waters represented. Namely, that this phenomenon did indeed take place in the pre-Big-bang era, testifying that time also existed.

Biblically, the Big-bang took place on the second day, known by the phrase "the separation of the waters from the waters". To conform to science, the Universal Cosmological Model also took the Big-bang as the zero instant of our Universe. This is due to the fact that the Universe, having undergone deflation during the period before the Big-bang, found itself condensed to a point where time ceased to exist.

It was his Big-bang that gave a new start to our space-time, or the Universe according to Big-bang.

1.13. For the common man, before the Big-bang, there was nothing. What is the scientific conception of "nothing" used in this statement?

For ordinary people, before the Big-bang, there was something; that is what caused the Big-bang. It was the scientists who said that before the Big-bang, there was nothing. The scientific conception of " nothingness " in this statement is that we cannot even talk about the Pre-Big-bang, because space and time being confused, everything started with the Big-bang.

Since I am also a common man, therefore, I believe that there was indeed a Pre-Big-bang. Therefore, your question deserves to be rephrased. I watch several scientific documentaries. Ordinary people always ask what was there before and scientists always answer "nothing". So, to quench their thirst, should we resort to metaphysics or to the first cosmological book par excellence, namely the Bible ? That is indeed what I did.

In my books, part of the series "The Verses of the Scholars", I have made a study on "Cosmology according to the Biblical and ancient model", which is also the title of Volume I. The result was that the Universe had a beginning, characterised by what can be scientifically called quantum vacuum or quantum nothingness.

It was Prof. Lawrence Krauss of Arizona State University who put forward the theory of quantum nothingness. While claiming that there was nothing before the Big-bang, which by the way answers your question, his theory is based on so- called virtual particles that are born and die in empty space in fractions of a second before the Big-bang.

However, according to the Universal Cosmological Model, this interaction between the particles led to cosmic deflation, which resulted in an extremely condensed Universe, and it was after the Big-bang intervened. Thus, the nothingness to which science refers corresponds to a quantum vacuum with virtual particles.

1.14. What is the evidence that allows anyone to

assert that the Big-bang created the Universe?

There is no evidence to assert that the Big-bang created the Universe. It was a transitional stage in the Universe. On the other hand, there is scientific evidence to show that the Big-bang did take place. While scientists explain this evidence in complicated terms such as the cosmological background, the Bible's story of creation as quire scientifically reconstructed in the "Universal Cosmological Model" has made the Big-bang demonstration in the books of the "Verses of the Scholars" series very simple.

To describe the Big-bang, the Bible first used the phrase "the separation of the waters from the waters" (Gen 1:6), demonstrating that we are dealing with a single mass that separates. The Bible speaks of the extent and science of expansion. With the waters below, the planets were formed. With the waters above, the light fixtures were formed. All of these resulted from something unique that the Bible calls water. This is why the elements found on earth are the same as those found on all the planets and exoplanets, and the same as those found in the stars (including the Sun) in the form of plasmas or fluids. This demonstrates the inescapable proof that there was a Big-bang.

1.15. What would you say to someone who does not believe in the Big-bang theory?

I would say to him that he is not wrong in so far as astrophysicists explain it in complicated scientific terms which are not always understood by common mortals. On the other hand, I would like to help him to believe in this theory with the following comments.

First of all, the Big-bang theory postulates that the Universe began with an original atom, according to Father Georges Lemaitre, before expanding. If this is the case, the material found in planets should also be the same in stars.

The story of creation according to the Bible, as scientifically reconstructed in the Universal Cosmological Model, has made this demonstration of the Big- bang very simple. With the waters below, the planets were formed. With the waters above, the luminaries were formed. All of this was based on something unique that the Bible calls water (Gen 1:6-14). This is why the elements found on earth are the same as those found on all the planets and exoplanets. They are also the same as those found in the stars (including the Sun) in the form of plasmas or fluids.

As to how the Big Bang and beyond came about, everything is developed in the books in the series "The Verses of the Scholars".

1.16. Do you think that the Big-bang theory is a satisfactory answer to the origin of the Universe?

The Big-bang theory is not a satisfactory answer to the origin of the Universe. Why is this? Because the Big-bang was only a transitional stage in the history of the Universe. The true history of the Universe that gives a satisfactory answer to the origin of the Universe is the restitution that has just been made in the Universal Cosmological Model, which restores the history of creation according to the Bible. This model is developed in the books of the series "The Verses of the Scholars".

In these books, the history of the Universe from before the Big Bang to mankind has been traced back in Volume I through a new theory called "The Universal Cosmological Model".

1.17. Why is the Big-bang theory taught, if it is only one theory among others?

The Big-bang theory is taught because it is justified by all the phenomena that we observe in the post-Big-bang era. Moreover, it is biblical and the author of the books in the series "The Verses of the Scholars" is the first to observe it or to make the world discover it. Soon the theories of the "Universal Cosmological Model", which scientifically reconstructs the biblical story of creation, will replace it.

In fact, the Universal Cosmological Model swallows up all cosmological theories, putting each one back where it should have been, in order to arrive at the theory of the Everything.

1.18. How can the Big-bang theory be proved beyond doubt?

Proof comes to us from the elements found everywhere in the Universe. Indeed, the Big-bang theory postulates that the Universe began with an original atom, according to Father Georges Lemaitre, before expanding. If this is the case, the materials found in planets should also be the same in stars. This is indeed what astrophysicists have been able to discover.

1.19. Why was the Big-bang theory so controversial in the beginning?

The Big-bang theory was so controversial in the beginning because it came from a religious man, the Belgian Physicist Father Georges Lemaitre. He was even opposed by Albert Einstein who criticized him for mixing science and religion. And his main detractor, the scientist Fred Hoyle, to make fun of his original atom theory, called it a Big-bang.

Throughout the 20th century, observations proved Georges Lemaitre right. And at the beginning of the 21st century, 100 years later, the research done by the author of the series of books "Les Verses des Savants" (The Verses of the Scientists), proved that it was really a theory whose origin is biblical. Consequently, Pope Pius XII was indeed right to congratulate Father Georges Lemaitre, despite the fact that the latter refused to do so because he did not want to be frowned upon by scientists.

1.20. Do you believe in creation or in the Big Bang?

Creation and the Big-bang are two different phenomena. Who says creation, sees the beginning of the Universe, but the Big-bang is a transitory stage in the evolution of the Universe. According to the Universal Cosmological Model, the Big-bang marked the end of a process, namely the process before the Big-bang.

The description of what happened before the Big-bang is reported in the books of the series "The Verses of the Scholars".

For a short summary, in these books, the author has made a study on "Cosmology according to the Biblical and Ancient Model" which is also the title of Volume I. This study revealed that the creation which was characterized by a vacuum at the beginning, called by default quantum vacuum, experienced deflation on the original particles until its condensation. It was after this extreme condensation that the Big-bang occurred; the phenomena that followed are described in the Standard Model in part and in the Universal Cosmological Model in full.

1.21. Does the Big-bang imply that the Universe has had a beginning?

The Big-bang really implies that the Universe has had a beginning. We are told scientifically that at the Big-bang, the Universe was close to 10 exponents 15°C. One cannot reach such a temperature simultaneously. This is to say that there was something in the past of the Big-bang which is called the pre-Big-bang described in the books of the series "The Verses of the Scholars».

1.22. What happened before the Big-bang?

You cannot get this answer in today's science. But in the

Universal Cosmological Model, the creator of this Model, who is none other than myself, has traced the situation from before the Big-bang to man. This Model makes a restitution of the biblical pattern of creation.

In this Model, as developed in the books of the series "The Verses of the Scholars", we are told that the Universe began with chaos, a vacuum, as described biblically in the words "the earth was formless and empty" (Gen 1:2). Something that is formless and empty is something that is materially non-existent to the naked eye. It is like a colorless and odorless gas in space. It is in this emptiness that the creator of the model demonstrated how things happened until the Big-bang was achieved, using quantum mechanics.

1.23. How can the Big-bang be a credible theory if the explosion cannot come from "nothing"?

Soon this question will no longer arise because the theories dealing with this question have just been solved through the Universal Cosmological Model.

For a short summary, the only book that talks about the phenomena that took place from before the Big-bang until we ended up with a condensed and hot Universe is the Bible. And then the Big-bang happened. I know that you have all opened your eyes wide, but the Universal Cosmological Model, developed in the books of the "Verses of the Scholars" series, has just made a scientific reconstruction of the biblical creation story and clearly reveals what was hidden during these periods.

1.24. Is the emptiness of space "nothing" (as it was before the Big Bang)?

According to the book Cosmology according to the biblical and ancient models, the pre-Big-bang is a reality that began with a vacuum. The following verse states : "In the beginning God created the heavens and the earth" (Gen

1:1). To make it clear that this was not our present earth, the Bible says : "The earth was formless and empty". Something formless and empty simply means something that does not have a material existence, i.e. the Universe at its beginning.

So, according to the above book, which scientifically reconstructs the biblical creation story, everything begins with a vacuum. The pre-Big Bang, which is said never to have existed, was a very eventful period before the Big Bang and the post-Big Bang. The author of the book is the first to discover these phenomena in the Bible.

1.25. Stephen Hawking said that there was nothing before the Big-bang. How can something form out of nothing?

The scientist Stephen Hawking, whom I respect very much, was a great physicist. However, a physicist can only refer to what is mathematically proven. Or he can either predict what is not yet proven.

But there is a question we must ask ourselves : what was the first cosmological book that actually spoke of the origin of the Universe? It is the Bible. The biblical conception of the creation of the Universe has been thrown in the dustbin by everyone. At present, this conception has just been resurrected through the books in the series "The Verses of the Scholars" by its author, who is none other than me.

The results that I have discovered without them knowing it before are :

– The Pre-Big-bang is biblical and was a very eventful period.

– The Famous Big-bang itself is also biblical.

– And after the Big-bang followed the post-Big-bang until the formation of the integral man.

The biblical scheme of the creation of the Universe, as given in the Universal Cosmological Model, is more logical and coherent than all the models. Our present scholars have not yet reached the level of Moses, the author of the biblical scheme of creation. They must make efforts and surround themselves with humility to advance in the sphere of the Pre-Big-bang.

1.26. It is said that the Universe appeared from a very small point responsible for the Big-bang. So what was the period before the Big-bang or pre-Big-bang like?

According to the Universal Cosmological Model, which is the restitution of the biblical scheme of creation, the pre-Big-bang period was an energetic period, as taught in quantum mechanics, although the authors are not aware that they described what was happening in the pre-Big-bang period. This is how the laws of quantum mechanics do not fit with general relativity. It was a very turbulent period before we arrived at the very small point responsible for the Big-bang.

What happened during this pre-Big-bang period is described in the books in the series "The Verses of the Scientists" dealing with science.

Section 2 : Gravitation

2.1. Why does the force of gravity exist?

First of all, gravity creates the cohesion of large structures. In addition, in the face of the pressure of the acceleration of the expansion of the universe, gravity prevents destruction, keeping the clusters together. If you had asked the question how gravity is created, I could have answered by referring you to Volume II of my book, in the chapter concerning "the universe of atheists". But by asking why it exists, there you are sending me to enter into religion because science has no answers and does not deal with why in relation to natural phenomena. All that scientists know is the manifestation of gravity or its effects.

2.2 How do you create gravity in space?

Gravity is part of the property of bodies, so it cannot be created as a thing has been created. In other words, it is part of the property of material. But through the equation $E=mc^2$ (E=mc squared), we see that the origin of material goes back to energy, as demonstrated in the Universal book, in the chapter concerning "the universe of atheists".

Where there is no material, there is still energy. Therefore, gravity in space is in an energetic state (this is a new notion developed in the above-mentioned book). In order to create it, that is to say to make it manifest, energy must condense through different processes, as quantum mechanics teaches (we understand that these are great notions). Once we have passed through the stage of virtual energy to real energy and have reached the stage of material, then gravity begins to manifest itself externally.

But this gravity was already present in an energetic way in space.

2.3. Why don't all the stars come together at one point because of universal gravitation ?

You asked your question as if the gravitation came from the Universe. But what I can say is that it is rather the bodies inside the Universe that possess gravitation. The gravitational power of each body is proportional to its mass, that is to say, it is limiting. Let's take the case of a magnet, it can attract metals that are close to it, but it cannot attract what is far away. Therefore, since the bodies in the Universe are separated by great chasms, it is difficult for all of them to merge into one.

In addition, there is dark energy, which is a repulsive force that, making the galaxies move away from each other, counteracts the gravitational force.

Section 3 : The Universe

3.1. Would it have been possible to understand the world and how it works without going through mathematics ?

In the book "Cosmology according to the biblical and ancient models" we are described how, in ancient times, the philosopher Empedocles (5th century BC) was able to describe the functioning of the Universe starting from the pre-Big- bang, through the Big-bang, describing how structures were formed in the post- Big-bang period up to the Big-crunch. I wonder whether he was simply a mathematician in his time or a simple philosopher.

There is also Moses, whose biblical scheme of creation we have reconstructed in the "Universal Cosmological Model" : how did he know how to make the division of days in the Bible without exploiting the study of fossils, and how did he discover that man is the last of the creatures ?

In other words, it is possible to understand the world and how it works as a whole without using mathematics. On the other hand, mathematics is important when you start to make analyses to understand the results.

3.2. What are the strangest astronomical objects ?

The air we breathe. Without knowing it, it is this air that connects us directly to the Universe, making us one with it. We ourselves are made of the dust of the universe.

3.3. Is there only one Universe ?

Referring to the first cosmological book par excellence, the Bible, we can say that there is only one Universe. This creation story has been masterfully reconstructed in a scientific manner in the Universal Cosmological Model, a model that is developed in the books of the series "The Verses of the Scholars".

According to its outline, the Universe had a beginning called the pre-Big- bang, experienced the Big-bang, and now finds itself in the post-Big-bang. It should be noted that the author of the series is the first to discover these phenomena in the Bible.

3.4. What event could affect the whole of the Milky Way?

Astronomers tell us that in 5 billion years the Sun will exhaust all its fuel and its disc will engulf all the planets. This is also one of the events that could affect the whole Milky Way. But according to other sources, in 4.5 billion years we are told that there will be a collision between our Galaxy and the Andromeda Galaxy. If we subtract this, we get a difference of 500,000 million years, i.e. we fall back to the moment when the countdown for the destruction of the Sun should already have begun. This means that earthlings may no longer be there to witness the destructive spectacle.

However, these predictions contradict the following verses :

Ecl 3:14 *"I have recognized that everything God does will last forever, that there is nothing to add to it and nothing to take away from it, and that God does this so that we may fear him".*

Eccl 1:4 *"One generation goes away, another comes, and the earth continues for ever. "*

Even if one is an atheist or a non-believer, could one not thank such a God who has created constants that science has not yet discovered for the fulfilment of these verses ?

These verses are not fulfilled by crossing our arms. God created man as the Lord of the Universe. The intelligence he endowed him with will enable him to discover solutions to these threats of massive destruction. This is indeed what NASA is doing in collaboration with other space centers. We should be very grateful for the work that astrophysicists are doing to safeguard our planet because this is also part of man's responsibility.

3.5. What are your favorite hypotheses and theories about space?

When I talk about space, I see the Universe. Sometimes the notion is confused with the common man. My favorite theories are those developed in the Universal Cosmological Model through the books of the series "The Verses of the Scholars", where the Universe had to go through three phases : the pre-Big-bang phase characterized by cosmic deflation, the condensed Universe and the post-Big-bang phase characterized by cosmic inflation. These three notions are taken from the Bible and scientifically reconstructed in the above books.

3.6. What scientific proof do we have of the existence of parallel worlds?

Some astrophysicists really believe that they have found evidence of a neighboring universe interfering with ours. This is the case of Laura Mersini- Houghton, a physicist at the University of North Carolina.

Once again, to support her evidence she resorts to the cosmological diffuse background.

Indeed, in 1992, the American satellite COBE detected minute temperature variations of the order of a few millionths of a degree in the 3 K temperature of the cosmic microwave background. These variations are interpreted by astronomers as 'density contrasts' : some photons were emitted from slightly denser - and therefore warmer - regions than others, as if they were hatching points for future galaxies! (Cf. : Sky&Space. December 2002, The universe is flat, by Azar Khalatbari ; P.37).

For the physicist Laura Mersini-Hougton, this difference

in temperature, which "one of these spots cover about 10 degrees of the sky", testifies to the presence of other universes that have been created and are similar to ours.

Note that these are also the same spots that are interpreted as the pre-Big- bang signature by some astrophysicists. As we can see, the debate is fierce, so much so that the reading of the diffuse cosmological background risks becoming a specialization in astrophysics given what is at stake. It is the only mirror that reflects back to us the image of what is real, but which we cannot see.

The reading of the diffuse cosmological background by Laura Mersini- Houghton, a physicist at the University of North Carolina, is not supported by the majority of astrophysicists. So, for your question, I can answer that there is no scientific evidence yet to prove the existence of parallel worlds.

However, the Universal Cosmological Model is not against any research, because each one of them sometimes leads to another positive result that was not expected.

3.7. Is there evidence of a multiverse?

I have learnt that in astrophysics one does not directly answer no, because whoever puts forward an idea surely has the evidence to back it up. You first start with research to understand and put forward your own. At this stage, if I say no, people will tell me that my answer is based on biblical faith. It is this judgment that I avoid in the Universal Cosmological Model.

Thus, I prefer to bring in the arguments of those for and against before making my conclusion, as my readers are free to do.

The idea of multiverse, which sometimes goes hand in hand with the idea of parallel universes, has been gaining ground among astrophysicists in recent years. According to this model, our Universe is only a tiny bubble in the middle of others. This idea is currently supported by Matthew Kleban, Associate Professor of Physics at New York University. His defense is based on mathematical formulas supporting his theories.

In Volume II of the series of books "The Verses of the Scholars", we said that if parallel or multiverse universes existed, they would already end up merging with each other and self-destructing through the exchange of energy.

I was surprised when doing my research, to find that the same point of view is also shared by some astrophysicists. But Professor Matthew Kleban has another explanation if such a phenomenon were to occur, as we can read : "The result is a wave of energy injected into our own bubble (our Universe) that propagates through the universe in what we call a 'cosmic wake'. Many objects would be affected by this wake, but especially the radiation from the cosmic background.

Professor Matthew Kleban is looking for this wake in the image of the cosmic microwave background, because, he says, "it is the oldest and most distant radiation and has therefore had more time to be affected by this kind of event. "This proves that, in his opinion, this collision has already taken place in the past. However, the evolution of the Universe, which I am researching, shows no evidence of objects that have been affected in the past. Therefore, for your question, there is no evidence of a multiverse.

3.8. Why isn't the Universe white?

If you turn to the first cosmological book par excellence, namely the Bible, we are told that in the beginning "there was darkness covered the surface of the watery depths ". Therefore, the initial state of the Universe was darkness. Due to its expansion, starlight is unable to fill it. So, as soon as the Sun disappears from the sky, the part that it used to illuminate regains its state of darkness. And, if all the stars were also able to disappear, the Universe would return to the state predicted in the Bible, i.e. darkness.

3.9. Can there be two different theories of the world, both true and complete ?

I answer this question taking into account the Genesis story of the Universe. All the theories that are proposed to us today are only in their infancy. However, it is thanks to these theories that God can raise the equivocation today.

If you find two people arguing about you, one maintains that you were born by Caesarean section because you have a big head and the other, on the other hand, maintains that you were born normally, to whom will we give reason between the two ? Only your mother can decide this question. The same is true of the theories that are being proposed to us about the Genesis of the Universe at the moment. The only true theory is what scholars have rejected, namely the creation of the Universe according to the biblical scheme as given in the Universal Cosmological Model and developed in the books of the series "The Verses of the Scholars".

3.10. What does the observable edge of the Universe look like ?

The edge of the observable Universe looks like a dark horizon whose background and purpose can no longer be seen. It is like when we describe the horizon of a black hole and its surroundings outside the light disk.

3.11. Could dark material be an agglomeration of dark dwarfs from an ancient Universe? Jean-Pierre Luminet says that black dwarfs will decompose into radiation and neutrinos will disperse in the cosmos.

We must first discover this dark material on the basis of observations or exact mathematics before we can address the points that concern you. But until then, even in physics, the point concerning its existence is not yet unanimous. Moreover, I intend to offer my hypotheses on this question in Volume V which is in preparation and which will deal with subjects that are part of the observations but not yet resolved by science.

3.12. Which question has not yet been sufficiently answered by science?

The question about the origin of the Universe, but the Bible has answered it without anyone as yet knowing how to interpret it. That is what I have just done.

Indeed, as the diffuse cosmological background requires experts in astrophysics to interpret it, there are also, in the Bible, certain verses with a scientific connotation that pastors or theologians cannot interpret. This requires an expert in the interpretation of biblical scientific language. The only expert so far is the one who has just reconstructed the biblical creation story through the Universal Cosmological Model, that is to say myself.

It is in the books of the series "The Verses of the Scholars" that the origin of the Universe is widely developed and, consequently, the author is also the first self-taught researcher to discover that the pre-Big-bang and the Big-bang are biblical.

3.13. Is it useful to send men into space?

Just as knowledge of France interests all French people, so too is man called upon to understand the Universe as his country. If we were not interested, today many people would still be worshipping the Sun or the Moon as in the village of Amazonia and sacrificing their children as offerings to make it rain.

So the knowledge of all the corners of the Universe is also very important for mankind. In the book Cosmology according to the biblical and ancient models, the author highlighted man's responsibility towards the Universe when God created him. And, it emerges that man has this responsibility to know everything about the Universe and to work on it.

3.14. Does the Universe have an end and a beginning?

The word "end" that you use has two hypotheses. We don't know whether you are talking about the end as a frontier or the end of the existence of the Universe. Let me first address the first hypothesis. Today if we know that the earth is round it is thanks to the one who went around it and came back to the same place. And then the satellites arrived and took images confirming this. As for the Universe, for the moment there is no machine that can go around it or photograph it from outside so that we can have a vision of its extent.

Given that the speed at which galaxies are moving away from us is greater than our machines, perhaps one day, after thousands of years, a galaxy that had disappeared from the scope of our telescopes because of expansion will come, by reappearing, whether the Universe has an end.

In the second hypothesis, if your "end" refers to the end of the existence of the Universe, we will answer the following. So far, science has proposed three main scenarios.

– If the expansion becomes too great, it can break the entire force of gravity and the entire Universe will be destroyed, including our cells. But where will all this material go? They will be there without structure. It is us who will no longer be there.

– If the expansion loses its strength and gravitation gains strength, it can cause a contraction forcing all structures to merge and crack. This is the Big- Crunch or the second scenario of the end of the Universe. But where will all these fused materials go? They will be there without structure. It is us who will no longer be there.

– If the expansion balances with gravitation, with what is called cosmological constancy, then the Universe will reside forever. But don't gloat because the end of the Sun is predicted in 5 billion years, which will put an end to planet earth.

I said at the beginning that science has proposed three main scenarios, but in the Bible there is a fourth scenario that joins the third. We can read it in the following verse :

Eccl 3:14 *"I have recognized that everything God does will last forever, that there is nothing to add to it and nothing to take away from it, and that God does it so that we may fear him".*

Now we can approach the paradox of the beginning of the Universe. For science, no one can speculate about the Pre-Big-bang, and that the beginning would be the Big-bang because that is where time and space began. In the Universal Cosmological Model, on the other hand, which reconstructs the biblical story of creation, developed in the books of the series "The Verses of the Scholars" dealing with science, we find the Pre-Big-bang, that is to say where the Universe really began and the cycles it went through until it reached the Big-bang, which was only a transitional stage, before beginning the after Big-bang.

3.15. The history of the Earth covers approximately 4.5 billion years. Are these light years or earth years?

The 4.5 billion years of Earth's history are Earth years, not light years. We talk about light years when we look back at the distance from one point to another. The universe being so large that if we have to translate all distances in km, we will have numbers with more than fifty digits.

For example, among the more than one billion stars already discovered by large telescopes and satellites, the closest one, Alpha Centauri, is far from the earth at an average distance of 40,398,000,000,000,000 kilometers (40,398 billion km). Thus, to evaluate such a distance, astronomers have taken as a reference the speed of light in vacuum which is about 300,000 km/s. Multiplying this by one year, we will have :

1 al = 300.000 km/s x 60 (1 minute) x 60 (1 hour) x 24 (1 day) x 365 (1 year) = 9.460.800.000.000 km or about 9.461 billion kilometers. So a light-year is equal to 9,461 billion kilometers.

Thus, to express the distance of a distant star, we divide this distance by

9.461 billion kilometers, and we find in its distance which is expressed in light- year. But when it comes to chronological values, it has nothing to do with light years.

3.16. If all the material in the universe comes from the Big Bang, how can we explain the diversity of matter found in the universe?

It is through this question that will know effectively that

the Big-bang was a reality. The diversity of material which one finds in the universe started only with only one mass qualified by the Bible of the "waters". At the time of the Big-bang or the separation of the waters from the waters, a part formed the planets and another part the luminaries. This is to say that the same materials found in the planets are the same materials found in the luminaries.

But where did the diversity come from? Everything starts from nucleosynthesis, that is to say that it is from hydrogen, which means what generates water, that all matter was created. We have explained the processes in the book.

3.17. What is greater than the Universe?

Astrophysicists have created two concepts for the universe. When they talk about an observable universe, they use a capital "U" which is our Universe. Therefore, the Universe with a capital "U" is part of the visible part of the universe and extends up to 45 billion light-years. This is the distance from the earth to the farthest star in the universe.

Behind this star, there is only a deep and dark void. And, if we include this deep and dark void, which we can no longer see, then we are talking about the universe with a tiny "u". In conclusion, we can say that "the Universe" is included in "the universe" and on the physical level, nothing is bigger than the universe.

3.18. Why is the Universe not a black hole? Move the Universe backwards and watch it become denser and denser. Forget the possible singularity, at some point it must be so dense that it forms a black hole. Why didn't it?

Your question is pertinent, but you confuse the black hole with the deformation of the fabric of space-time. Indeed, all deformations of the fabric of space-time are not to be confused with a black hole. The black hole just like the mass creates this deformation.

Therefore, by moving the Universe backwards, we find ourselves to count to a very dense mass which will create the deformation of the fabric of space-time. However, you are also right to speak of a black hole because we do not yet know how this mass behaved during the Big Bang.

For some, the Big bang was an explosion, i.e. the universe threw up its mass like a black hole would. In this case, one will be able to say that the Universe came out of a black hole. Other scientists say that it was only an expansion as we see it today. In this case, we can say that it was only a distortion of space-time.

No one will be able to assert either hypothesis with confidence.

Section 4 : Religion

4.1 As a believer why do you think there is a God?

For a believer, the fact that God exists is not part of thought but of reality. The universe was created more than 13.8 billion years ago, whereas Moses, who wrote the first cosmological book, lived only 3,500 years ago.

Who inspired him to write an almost perfect scheme of the creation of the Universe as depicted in the "Universal Cosmological Model"? This Model is the first of its kind to clarify the biblical creation story and therefore proves that God is not just a theory, but a dynamic Being.

4.2. In which books is the name "God" not mentioned?

It used to be in the books of mathematics. But the books in the series "The Verses of the Scholars" have just put an end to this. With these books, which mathematically reconstruct the biblical creation story, God has just made his resounding entry into the exact science since the creation of the Universe. And this is the work of an African.

4.3. Did God create the Big-bang?

First of all, the Big-bang is a transitional stage in the history of the Universe. According to the Universal Cosmological Model, a model that reconstructs the biblical creation story, the Big-bang marked the end of a process, namely the pre- Big-bang process. The description of what happened before the Big-bang is described in the books of the series "The Verses of the Scholars".

For a short summary, the Big-bang is a long way from what God actually created, namely the Universe at its beginning. It is after several processes, as described in the above mentioned books, that the Big-bang came into play. Was it in God's plan or not?

Let us take the case of a mango tree; who created the trunk, the branch, the leaves and later many mangoes in the same tree? It is simply from a small nucleus that was planted on the earth that all the processes originated.

God was not there to say that the Big-bang would happen, that this would happen, etc.... All his will of what would happen to the Universe was already inscribed in the first particles as in the example of the nucleus. And as it was part of his will for it to come to man, we can conclude that this God who created the Big-bang and all the phenomena that followed until the formation of man.

Another example with regard to the Universe : when you plant a baobab seed that you have carried in your pocket. Later you will have a big tree with leaves and maybe you won't even know how to climb the tree in question. Was it you who put the branches and leaves in? Tool was already inscribed in the small seed. The same goes for the Universe.

The Rev. Dr. Sun Myung Moon, who is the only great religious thinker on the origins of the universe, said this and we agree with his point of view :

"If you enter a mystical state, you will understand that a single grain of sand contains all the principles of the universe. You will see even in a single atom the infinite and inexhaustible harmony of the universe. Despite our limited understanding, we cannot deny that a complex network of forces is at the origin of all existence. Molecules contain atoms and the atoms of elementary particles. Yet even these entities are not devoid of consciousness. Each is endowed with a consciousness and a specific purpose". (Cfr : Cheon Seong Gyeong, FFPMU, p.85)

According to the Universal Cosmological Model, which is the only model to trace the situation from Pre-Big-bang to after Big-bang to the living, the Big- bang theory emanates from the theory of the original seed. So when we say that God created, we are referring to the seed that created many structures.

4.4. What are the miracles that science cannot explain?

First of all, a miracle is something that cannot be explained scientifically. Scientists like to talk about an enigma because the word miracle has become much more religious. However, as science progresses, several enigmas are being elucidated. But where science will one day come up against is at the origin of energy that has been transformed into virtual particles.

4.5. How could I learn the Principle of Creation for my general culture without going to a Moon Church center?

The Church of Unification, now called "The Holy Community of the Heavenly Fathers", has enough teachers who can give you, free of charge, according to the schedule agreed upon at your home or office, the teachings of the Divine Principle, which the Principle of Creation is the subject of the first chapter. You can get in touch either by going to a local center or by using the Google search engine by typing "FFPMU/ (THE NAME OF YOUR COUNTRY)". Once you have entered, you can have all the information; telephone number, address. If your local center does not have a website, you can contact FFPMU/France who can direct you.

4.6. How much credit can we give to the Universal Cosmological Model compared to other models?

Its logic from before the Big-bang to the formation of men and women; this is what is missing from other models which are only the fruits of human thoughts based on effects. Such a pattern can only emanate from the one that religion presents as the Creator of the cosmos, namely God. I was only a discovery of this pattern and I make it discoverable through the Universal Cosmological Model.

This is to say that my bibliography has no influence, nor the continent on which this model was created. This model is part of the ancient wisdom of the Hebrew peoples known in these days as the Children of Israel. Two parallel ancient models not derived from religion have confirmed and completed it; the ancient western model developed by the Greek philosopher Empedocles in the 5th century B.C. and the oriental model which dates back to the 12th century B.C. as I developed them in Volume I.

In our time, its confirmation began around the 1920s when it was discovered what happened on the second day of biblical creation, i.e. the Big-bang or "the separation of the waters from the waters" (Gen 1:6). And, it was from 1960 onwards that those who began to think about the pre-Big-bang era came to the conclusion that there was an abyss as the Bible says which science calls these days of collapse or black hole, before the emergence of our Universe. And, it was in 2020, a hundred years after the discovery of what happened on the second day, that the true horizon of the pre-Big-bang or what happened on the first biblical day was revealed through the Universal Cosmological Model.

Thus, the Universal Cosmological Model, compared to other models, is not the result of human thoughts. Rather, it is a discovery that the author came to present to the whole world in order to allow all specialists to access it and correct their thoughts by drawing inspiration from the writings of the first cosmological scheme of the universe contained in the Bible. This is the injunction that God ordered me to tell everyone.

Let's just wait for the scientists to get started and in a short time the confirmations will begin to fall like raindrops, because God himself is looking into it.

Section 5 : Miscellaneous

5.1. Biblically, whenever there was a plague in the Old Testament, God would eventually intervene. Currently, the coronavirus pandemic is plaguing the whole of humanity. Is this a divine punishment? Is there no longer a Moses in our time?

Biblically, God created in 6 periods and all that he created is good if each thing keeps its place and is used only for good. The snake's venom is good for its survival. But if it leaves the snake's body and enters another body, it becomes destructive and the snake is aware of this. The same is true for viruses and bacteria. They have their lodgings where they act positively on their natural carriers.

Moses in the Bible was aware of their destructive power and dictated many of the barrier gestures so that such germs for humans would not come and decimate them in the camp of Israel during the exodus.

Let's first talk about this coronavirus that is decimating humanity at this time and that cannot leave indifferent all those who are concerned about the future of humanity.

According to a study by the National Institute of Health (NIH), researchers had come to the conclusion that the coronavirus is transmitted by air, that is to say when an infected individual coughs or sneezes within one meter of another individual without blocking it with his elbow. Even so, scientists have identified that the coronavirus, although it remains "fragile", has a certain lifespan in an outdoor environment. According to the results of the study, the Covid-19 virus would be resistant for 3 hours of time in the air (Cfr : https://parismatch.be/actualites/sante/381736/coronavirus

-combien-de-temps-le- virus-survit-il
-dans-lair-et-sur-les-surfaces,
Published on March 17, 2020).

If in a given environment; open or closed, many people emit coronavirus particles, they form as a permanent microbe or those that die are replaced by new ones that are emitted. Thus, whoever passes by with a fragile health, risks being contaminated. Even if you think that you were alone without realizing that 30 minutes before you, someone emitted the virus particles. This is what is currently happening in the large cities of the West.

Preventive measures are more important than curative measures. At our level, we can only talk about curative measures, because what has been dictated so far will not be able to stop this pandemic. While waiting for God to inspire researchers on curative measures, we want to act as Moses did, that is to say, by focusing our contribution on barrier gestures.

Certainly, when the people of Israel were crossing the desert, Moses dictated several barrier gestures that we developed in Volume V of the collection of books "The Verses of the Learned" entitled "From Noah to Barack Obama". There, we revealed the providential causes of this pandemic whose duration should restore the time of the flood. The real hopes should begin in the month of March 2021, referring to our research and not to our prophecy, when Noah came out of the ark and offered the offering to God. Namely that Noah remained in confinement for a year and only came out on a certain February 27th of the year and offered the offering.

In total unpreparedness, we could not respect the same time of confinement.

All this is developed in Volume V, I had said. As you read Volume V, you will understand God's concern for man and that all punishments are created by ourselves.

Returning to the barrier gestures dictated by Moses in the Bible during the Exodus, one of them was that the people should wash their hands at all times. Moses even listed it as an impurity (sin) to eat without washing one's hands. Jesus understood that this was only a barrier and banished it from sin by saying that "it is not what goes into the mouth that defiles a man" (Mt 15:11). However, this did not mean that we should lower our guard against this divine commandment and not consider it as if it was only part of the Old Testament.

With the Coronavirus, the main barrier gesture of bringing masks is no longer enough to feel protected. We have said, according to researchers, that the particles of several viruses, including coronavirus, float in the air and we cannot see them with the naked eye. And, when a microbe floats in the air, there is only the body's resistance left to deal with it.

The mask not only makes us uncomfortable, but does not provide a 100% barrier. Just when we think we are alone and safe, the virus may be there, floating. The mask does not repel the virus. The invisible must be chased away by the invisible. This is our great contribution that we want to develop in the following paragraphs that we call "The Phanzu solution's".

In addition to the barrier gestures dictated by the health authorities, I offer my design for free to industrialists and businessmen who can materialize them. The Bible says "you have received freely, given freely". These are ideas that can be given for free and not material things that require investment.

The mask can be replaced by the kepi fitted with a mini ventilator turned downwards. The vertically descending wind will prevent any invisible particles in the air from reaching your nose or lips. In addition, if the wind is strong enough, it may even send particles of microbes or droplets of saliva that we emit when we speak directly downwards. By coughing abruptly, if we haven't had time to put the elbow, the particles will suffer a first shock from these winds and consequently, will take little time to float in the air. This will also protect others. This is more convenient than the mask and more reassuring when you are alone or walking.

Bringing such kepi or fan hats with a red pilot light to show that it works will make our relationships normal. We can kiss each other by touching only the tips of the kepi or hats.

These arguments lead us to add that in all enclosed areas or areas with a large number of public visitors, the use of ventilators to expel air can be preferred to the use of air conditioners. Let this be extended to trains, vehicles and offices. Let's stop the spread of the corona virus by talking about this concept, which I will come back to in Volume V to industrialists who don't have time to read books. This is the design of the hat or portable antivirus device on the forehead or head called "The Phanzu solution's".

My daughter LEMA PHANZU NERIA wearing a kepi form with electronic barrier gesture.

The propeller winds prevent all airborne microbes (bacteria and viruses) from entering the eyes, nose or mouth. Any firm that would manufacture some that it would not forget the inscription " The Phanzu solution's". In addition, with ceiling lights, any airborne microbes emitted are driven out and the wind coming out through the vents with more power than the outside air will prevent others from entering.

People drop their cumbersome masks when they feel alone in an environment and become infected. But with the Phanzu solution's, you don't need to stop the armature and the wind emitted by the propellers chases away any airborne microbes by preventing them from touching your eyes, nose and mouth. Moreover, it is not cumbersome. Even children will now be comfortable.

In Africa it is said : a fetishist always keeps an ingredient for himself so that wherever he is, his product keeps its originality compared to others.

CHAPTER VIII :
Science and Religion

– Tim 2:3-4 *"This is good and acceptable in the sight of God our Savior, who wants all men to be saved and to come to the knowledge of the truth".*

Section 1 : PROPHETIC MESSAGES TO SCHOLARS

J.B.S. Haldane said : "*The progress of scientific knowledge does not seem to take away a shred of mystery from our Universe nor from our inner life within it.*"(Cf. : J.Baker : 50 clés pour comprendre l'astronomie, ed. Dunod, Paris 2016, p.42).

1. The unity between science and religion

God's ultimate goal is for science and religion to work together. It is like body and mind. When science denies religion, it is like the body denying the spirit, and yet they are united as long as a person is still alive. So, we will have to complement each other, we will have to agree that science and religion are united. Jesus spoke to the scientists when he said to his contemporaries :

Jn 3:12" *If you do not believe when I tell you about earthly things, how will you believe when I tell you about heavenly things ?* "

I am not the one sent to speak of heavenly things, because it is in the heart of every scholar, putting God in the first place, that this will be revealed. Wasn't it in a church that Mr. Edwin Powell Hubble was inspired, and with the observations of the Belgian Catholic canon Georges Lemaître, both astronomers, they came up with the Big-bang ? Except that the latter, who was a priest, was unable to look in the Bible to discover when, in the writings of Moses, the Big-bang took place. I am the only one who did and discovered it.

In the history of ancient religion, scholars called the prophets "the prophets". They preferred this name for two reasons : to believe that their calculations and discoveries were divine revelations and, it was an honor to distinguish themselves from others by having revelations from a powerful and invisible God.

In the Bible, the story of people like Noah, Elijah, Joseph, etc., whose prowess I revealed in Volume IV of the series of books "The Verses of the Scholars", effectively proves that they were scholars on the same level as our present day scholars, although Moses remains unbeatable.

Jesus Christ himself had told his contemporaries that they could not assimilate the truths and that this would come later through another prophet whom he called the Spirit of Truth, because God does not see the flesh but the spirit that will be inside that prophet. Let us read about it :

Jn 16:12" *I still have many things to tell you, but you can't bear them now. 13 When the Spirit of truth comes, he will guide you into all the truth. For he will not speak on his own, but he will speak whatever he hears. He will also declare to you what is to come ".*

And Saint Paul went even further, calling his generation little children in of knowledge of the materials and that they were not able to have difficult things explained to them. This is how he concluded by saying that when true knowledge comes, everything that has been taught up to that point will disappear, as we can read :

1Co 13:9 *"For we know in part, and prophesy in part, 10 but when that which is perfect has come, that which is partial will disappear".*

It is understandable why I am not afraid to say that all those who have not yet studied the teaching of the Divine Principle of Rev. Dr. S.M. Moon, are still two thousand years behind in their theological and philosophical thoughts, because the spirit of truth has already spoken. Therefore, in addition to what Jesus Christ said in describing his generation as low-minded (Jn 16:12), let us read what St. Paul completed :

Hey 5 : "*11 We have much to say on this, and things that are difficult to explain, because you have become slow to understand. 12 For you, who for a long time should have been teachers, still need to be taught the first rudiments of the oracles of God; you have come to need milk and not solid food. 13 Now whoever is in milk has no experience of the word of righteousness; for he is a child*".

The biblical story shows us that every two thousand years something special has always happened in providential history. Historians call this the stuttering of history. Therefore, scholars must be curious to listen to the message of the one who came after two thousand years of Christianity to shed light on the cause and effect.

The Divine Principle of the Moon couple is the parables revealed from Genesis to the Apocalypse to the providential history of our times, by God himself who wanted to reveal the truth about both spiritual and physical enigmas. . The Divine Principle does not list these mysteries, but it is through its understanding that we can solve different enigmas. It is like the scientific knowledge that enables people to create.

Thus, if science is still struggling to solve many enigmas, on the side of religion, everything is already revealed and explained by God through the Rev. Dr. S.M. Moon ; from the mystery of the beginning of the Universe, through the three great stages of life that every man is called to go through and this through the exact knowledge of the sin of Adam and Eve, to the process of resurrection, which even the passage of Jesus on earth did not leave a convincing answer to the Pharisees and Sadducees. As a result, Christians still maintain a false myth around this question. As Jesus said in Jn 16:13, he will also tell you things to come. Even explanations about the spiritual world are no longer a secret for these disciples.

All these clarifications are made to say that the Word of God can be completed from one sacred book to another. This is why a true exegete; scholar or researcher does not limit himself in his research because of his doctrine.

As a Catholic by birth, to discover for example from the spiritual to the Universe Pre-Big-bang that I cannot publish and from Pre-Big-bang to Big-bang, I had to crisscross several religious and esoteric movements and so many other Churches that publish their doctrines. I have crisscrossed them with a precise objective as a self-taught researcher to understand the mystery from the spiritual to the physical and not as an adherent.

In the end, I found that the teachings of these doctrines find their clarification in the Divine Principle of the Moon couple. At present, I am only devoting my research to cosmology since God appeared to me in 2012 when I discovered that the Big-bang is biblical, giving me the injunction "to tell everyone, especially scholars, that they must be inspired by one model, namely the biblical story of creation". All that remains, as I said, is only in its infancy.

At first I myself thought I was incapable of doing so, but after ten years of research I have come to the conclusion that the "biblical creation narrative" is the only true model. Being neither an astrophysicist nor a physicist, "The Universal Cosmological Model", as I develop it in the books of the series "Verses of the Scholars", is only an introduction and it is up to the scholars themselves to develop it.

Moreover, I invite all my brothers, the scholars and those who are in the Christian sciences and esoteric sectors, they now have within their reach a message that responds to their aspirations. A message that is given neither in the dark, nor in secret, nor in closed places, and which is not followed by any initiation. This message shows us the way to perfection (Mt 5:48), both spiritual and intellectual, so that we may become, not saints or teachers, but perfect persons, true children of God, who have their place in his Kingdom. This message is none other than the teachings that the Moon couple brought us, so that the saying that "perfection is not of this world" may cease.

GENERAL CONCLUSION

If the post-Big-bang era is characterized by the formation of structures in the Universe, the Bible has just revealed that the pre-Big-bang era was a sphere where only energies were in action with different phenomena ranging from the collapse to the bursting of light. All these phenomena are described in a pictorial language, on the first day of creation, which we have explained. This gave birth to the Universal Cosmological Model, which is the restitution of the biblical scheme of creation.

This restitution has shown that the biblical story of creation is real cosmological scheme, which has been confirmed by current discoveries, from Pre-Big-bang through string theory, through Big-bang, to after Big-bang as developed in the standard model.

Of course, we have not used a real scientific language, but when we discover a sphere and indicate where it is, it is to allow all the specialists to access it and to go and make all the adequate calculations. It wasn't up to the person who had discovered America to say how big it was and what treasures were hidden there.

Just as Georges Lemaitre showed us the Big-bang and Fred Hoyle came with nucleosynthesis, the entry into the pre-Big-bang sphere by scientists a century after the discovery of the Big-bang, will revolutionize many theories and bring us new ideas in cosmology.

Since mankind started asking questions about the Universe and answering them, several Models and schemes have been issued. On the basis of the observations, some have simply been disqualified. In spite of some observed imperfections, the Big-bang theory and the Standard Model have continued to impose themselves, but without being able to explain everything. However, the Universe was not designed by two or more Models. These models are only in their infancy. This is to say that in the end, there should only be one Model. If even two contradictory models persist, it is because we have not yet reached a true model.

As we have said, this single Model was already written on papyrus 1. 500 years BC. Being recorded in a sacred book called the Bible, no one has really been able to explain and exploit it. The beginning of the scientific restitution of this Model in the book "The Verses of the Scholars" begins to make the secret of creation clear.

Thus, the first author of this restitution in a scientific manner also became the first person who really discovered the secret of the Universe and called it "The Universal Cosmological Model". Therefore, all the theories and other models should be added to it to explain the different stages.

For lack of observation, the Big-bang was not unanimous until then. What I have just clarified in the writings that date back to around 1500 BC, means that to this day no one can doubt this phenomenon any longer. It is an earlier event that cannot be discovered through a telescope. The only certainty is the evidence that has come and gone and which we are witnessing today.

Without referring to scientific evidence, the Bible says that after the expansion of the original bubble called "waters", one part (the waters below) was used to form the non-bright objects (the planets) and the other part (the waters above) was used to form the luminaries (Ge 1:14), including the stars (including the Sun). These are the two great biblical proofs of the Big-bang.

If we have to go down into the secondary evidence and if the same bubble was used below for the formation of the earth and above for the formation of the luminaries, it is because the same elements should be found everywhere. Science has already proved this, even to the point of saying that man is the fruit of stardust. Man is defined as a microcosm in the Principle of Creation, for it is in him that a concentration of all the elements of the cosmos is discovered. Who can still stand up and say that the biblical creation story is a legend?

I have brought plausible arguments into this book. Everything I have written is more than the image of a telescope, because it is part of everyday observations. You don't need to be a physicist or an astrophysicist to understand this.

With the help of arguments from the Universal Cosmological Model, one can now, from high school and college onwards, teach the preliminary notions of cosmology to students, the Big Bang having now been unanimously accepted.

I come back to the divine recommendations which were dictated to me, namely : "Tell everyone to go back to the biblical scheme of creation", that is to say to base their thoughts on this model.

Thus, this conclusion closes for this volume the biblical situation from Pre- Big-bang, through Big-bang, to after Big-bang. Later, other astrophysicists, exegetes, philosophers, theologians, who will support the biblical models will make each point the subject of a book or a thesis.

Bibliography

– BOOKS

– Presentation of the Divine Principle : Culture et Paix, Paris, 2013.
– The Divine Principle, Level 4, Introduction
– Aurélien Barrau : Big-bang and beyond", Dunod, Paris 2019.
– Claude Sarrailh, Le secret de la vie, Librinova Publishing, Paris, 2018
– J.Baker : 50 Keys to Understanding Astronomy, ed. Dunod, Paris 2016
– Physicist Richard Feynman in "Lessons in Physics".
– Damian Grint, The Mysteries of the Universe, ed. Christophe Colomb, Paris, 1985
– Broglie (Louis, duc de), la Physique nouvelle et les Quanta, Paris, Flammarion, 1974
– Blandine Plucher, quantum physics for dummies, First edition, Paris 2018,
– Etienne-Klein : Speech on the origin of the universe ",-2010
– Stephen Hawkins, The Universe from Facts to Theories
– Stephen Hawking, A Brief History of Time, trans. I. Naddeo-Souriau, Paris, Flammarion, 1989
– Challenges of science-Le monde.fr, the Theory of Everything
The challenges of science ; the origin of life - edited by RBA Coleccionables,
S.A.U. Barcelona, 2018
– F. J. Dyson, Infinite in all Directions, Harper and Row, 1988
– Linda Gamlin, Les origines de la vie, ed. Artis-Historia, Brussels. 1989

– Gilles Dowek, Ce qui on ne peut parler il faut l'écrire, ed. Le Pommier, Paris, March 2019
– Ilya Prigogine : "Time to become".
Journey through the cosmos ; Emptiness and nothingness, edited by RBA Coleccionables, S.A.U. Barcelona, 2020

- ENCYCLOPEDIA

– "quantum chromodynamics." Microsoft® Encarta® 2009 [DVD].
Microsoft Corporation, 2008
– "Interstellar dust." Wikipedia, the free encyclopedia. July 27, 2016
– "Gravity." Microsoft® Encarta® 2009 [DVD
– Joe Bergeron, Encarta ® 2009 : Modelling a black hole
– The planets of the solar system ; encarta
– Elementaryparticles."Microsoft®Encarta®2009[DVD]. Microsoft Corporation, 2008
– Bing Bang Theory ; Microsoft ® Encarta ® 2009
– Wikipedia ; " Elementaryparticle
– Wikipedia : Great Unification
– Empédocle, Wikipedia
– Wikipedia ; Cosmological principle
– Chaos. "Microsoft® Encarta® 2009 [DVD]. Microsoft Corporation, 2008
– Wikipedia : GN-z11
– Olbers, Heinrich Wilhelm Matthaus "Microsoft® Encarta® 2009 [DVD]. Microsoft Corporation, 2008

– REVIEWS

– Pour la science, special report July 1997 Black holes
– Ciel&Espace magazine n°413, 10/2014
– Magazine Ciel&Espace n°423 - August 2005 ; The strange Universe of Mr HAWKING
– Magazine " Les secrets de l'Espace ", special issue n°7, the mysteries of the Universe, what happened before the Big-bang ?
– magazine ciel et espace, August 2005, what is the expansion of the Universe ? by Capucine Casati
– Ciel&Espace August 2005 ; Liquid water very early on earth
– Sky&Space. December 2002, The universe is flat, by Azar Khalatbari
– Revue La Recherche N° 551, P.34, 09/2019, Paris, Une nouvelle source d'information sur le cosmos, S.G.
– Ciel& espace, n°365 October 2000 Black holes, how to find them ?
– La force qui vient du vide - Astrid Lambrecht in monthly magazine n° 376
/06/2004,
– Science&vie Hors-Série n° 288, 09/2019 ; J.B. Veyrieras

– ARTICLES

– Bulletin de l'Union des Physiciens n°593 (April 1977) and n°577 (October 1975)
– H. GIÉ : Le moment cinétique à travers l'Univers* BULLETIN DE L'UNION DES PHYSICIENS P.427 N° 712, Paris

– WEB SITES

– http://www.matierevolution.fr/spip.php ?article38Le vide,... not so empty, le23/06/2010, parRobert Paris
– matierevolution.fr, Dark matter, dark energy : the missing link ?13/11/2007, by FaberSperber, Robert Paris
– http://www.matierevolution.fr/spip.php ?article38Les contradictions des quanta, 02/06/2007, by FaberSperber, Robert Paris
– http://www.matierevolution.fr/spip.php ?article43, Levidematerial destroyer/constructor,
01/07/2010, by FaberSperber, Robert Paris
– http://www.matierevolution.fr/spip.php ?08- Light and matter, laws from the void,Tuesday 14 July 2009, by Robert Paris
– http://www.matierevolution.fr/spip.php ?article, La symétrie brisée, 22/03/2009, by Robert Paris
– http://www.matierevolution.fr/spip.php ?article, Les bulles de vide et la matière, 16/05/2008, Robert Paris
– http://www.lalibre.be/archive/qu-y-avait-il-avant-le-big-bang ; published on 15/12/2005
by GUY DUPLAT
– trustmyscience.com/new-and-old-big-bang/ by Stéphanie Schmidt, 9 August 2019
– https ://trustmyscience.com/un-drou-supermassive-black-supermassive-denude- parcourt-l-University-)
– https://trustmyscience.com/reflexion-n1-lespace-temps
– trustmyscience.com/telescope-close-trace-history-stellar-stellar-in-the- world/
by Thomas Boisson,30 /11/ 2018
– directory-diggdes.coursgratuits.net/c7_bigbang/p715_partvirtuelles.php. byOlivier Esslinger

– astronomes.com/the-big-bang/virtual-particle/ on 9 August 2017 by Olivier Esslinger
– astropolis.fr/articles/history-of-the-world/the-zero-or-initial moment
– Dr. David P. Stern Mail to Dr.Stern:education [at] symbol phy6.org
– https://ingeniumcanada.org/innovation/fr/recit/un-trou-noir-lorigine-de- lunivers
– lestoutespremieresfois.com/1er-coup-daccelerateur-de-lexpansion-cosmique ; by Fabrice, Updated 8 May 2014
– astrocanada.ca/le_cosmos_contemporary_mater/Fra
– http//simulum.bio.uottawa.ca/bio2525/Notes/the-Protozoa
– http ://www.cosmovision.com/protozoaires.htm"
– https://www-n.oca.eu/vigouroux/chp1/chp123.html ;Âge du Soleil, Vigouroux Anne, Friday 13 September 1996
– DiagramofPenrose : Chaours.rv.pagesperso.orange.fr/physics/Stars/Black/penrose.htm
– vinsvignesvignerons.com/Geology-Geology/Une-breve-histoire-de-la-planete- terre
– http://www.elementschimiques.fr/ ?fr/decouverte/origine -de-la-matiere
– futura-sciences.com/sciences/univers-premiere image-black hole-supermassive, Laurent SACCO, 13/04/2019
– futura-sciences.com/sciences/definitions/physical
– JP LUMINET in Futurascience, l'infini de la matière : le vide, published on 25/01/2018
– Laurent SACCO,www.futurascience, Black hole
– https://www.futura-sciences.com/sciences/univers-energie-noire-4004
– futurascience.com/planets/definitions/botanical-cyanobacterium-122
– 10 essential questions about the Universe. J.P. Luminet, www.Futura-Sciences

– lestoutespremieresfois.com/1er-coup-daccelerateur-de-lexpansion-cosmique ; by Fabrice, Updated 8 May 2014
– sciencesetavenir.fr/qu-est-ce-que-la-matièree-noire-et-l-energie-noireBy Erwan Lecomte and AzarKhalatbari on 31.10.2018
– ces.org/articles/cosmology-fastoche-33
– zevenzeur.files.wordpress.com, the quantum vacuum
– zevengeur.wordpress.com/2014/01/16/history of the universePart 6 : Janus model, the future of cosmology
– zevengeur.wordpress.com/
la-chromo-dynamic-quantantic
– zevengeur.files.wordpress.com/2018/07/instant-zero
– zevengeur.wordpress.com/2014/02/11/The quantum vacuum : a new continent to explore
– sciencetonante.wordpress.com/2013/09/30/
the7wondersofquantum- mechanics

– SCIENTIFIC DOCUMENTARIES

– Where does the energy of the Big Bang come from?" by Hubert Reeves
– Aurélien Barrau; lecture : What was there before the Big-bang?
– New unified field theory : explanation of the Dynamic Universe
– David Louapre : "Quantum mechanics in 7 ideas
– New Unified Field Theory (Explanation of Gravitation)
– the origins of the earth"; written by Alice Harper and Nigel Henbest.
– Cosmos - Beyond the Big-bang"; written by Matt Hickey

– Peter Parks/Oxford Scientific Films

Sommaire

Librinova
www.librinova.com

www.ingramcontent.com/pod-product-compliance
Lightning Source LLC
LaVergne TN
LVHW052246200926
848481LV00002B/290